M. Cresti   S. Blackmore
J. L. van Went

Atlas
of Sexual Reproduction
in Flowering Plants

With 93 Figures and 101 Plates

Springer-Verlag Berlin Heidelberg GmbH

Prof. Dr. Mauro Cresti
Dipartimento di Biologia Ambientale
Università degli Studi di Siena
Via P. A. Mattioli 4
53100 Siena
Italy

Dr. Stephen Blackmore
Department of Botany
The Natural History Museum
Cromwell Road
London SW7 5BD
United Kingdom

Prof. Dr. Jacobus L. van Went
Department of Plant Cytology and Morphology
Wageningen Agricultural University
Arboretumlaan 4
6703 BD Wageningen
The Netherlands

ISBN 978-3-540-54904-8   ISBN 978-3-642-58122-9 (eBook)
DOI 10.1007/978-3-642-58122-9

© Springer-Verlag Berlin Heidelberg 1992
Originally published by Springer-Verlag Berlin Heidelberg New York in 1992

Production Editors: Susanne Fink and Renate Münzenmayer
Typesetting: Camera ready by author; printing: Beltz, Hemsbach; binding: Schäffer, Grünstadt
31/3111-5 4 3 2 1 – printed on acid-free paper

Foreword

From the dawn of time, man has appreciated that an ability to manipulate the reproductive phase of plant development has been key to the successful exploitation of crops. In more recent years, the cells involved in plant reproduction have taught us many of the general principles of developmental and cell biology - applicable to both plants and animals.

It is thus not surprising that, as the new technology of molecular biology comes of age and can be focused on "real" biological systems rather than model processes, interest in plant reproduction has intensified. However, as many an over-zealous molecular biologist has found to his or her cost, the cells and tissues involved in plant reproduction do not lend themselves well to traditional molecular analysis; typically floral parts - particularly the anthers and pistil - contain a bewildering array of tissue and cell types, all following very different developmental pathways. Further, the cells participating in gametogenesis itself are very few in number and develop very rapidly. Even were access to these cells straightforward - which it certainly is not - many key interactions occur in environments to which the methodology of molecular biology (which has usually been developed for animal systems) is completely unsuited.

Despite these technical problems, the potential scientific and commercial advantages of being able to manipulate plant reproductive systems remain vast. For example, an ability to control recombination, to regulate sexuality, inflorescence number and flowering period, and to be able to transfer systems of self-incompatibility to new species would transform plant breeding as it is known today. Cynics may say that this would only serve to add to food surpluses and to the profits of the plant breeding industry, but a moment's thought reveal that this is not the case. There is every chance that an ability to manipulate the plant's own gene manipulation apparatus will lead away from "fermenter farming" and towards "greener" agricultural and horticultural practices - particularly for the developing world.

An understanding of plant reproductive cell and molecular biology is thus of the highest priority. It has been recognised as such by the European Commission, which supports research in this area through the BRIDGE Initiative and the ERASMUS student interchange scheme. Collaboration under the ERASMUS scheme has resulted in the production of this Atlas, which not only makes a valuable contribution to our understanding of plant reproductive cell structure, but will also - through the striking images portrayed within it carry the excitement of working with these elegant and important cell systems to a new wider audience.

Hugh Dickinson

Oxford, September 1991

Acknowledgements

The concept of this volume emerged during collaboration between the Universities of Siena, Wageningen, Lyon and Reading within the framework of the ERASMUS programme of the Commission of the European Communities. The purpose of this collaboration was to develop a common basis of knowledge for the students of the four participating universities, and to facilitate the fruitful exchange of students. During the preparation of the volume, the Natural History Museum, London also became involved in the project. The authors particularly want to thank Prof. H.G. Dickinson of the Department of Botany, University of Reading, and Prof.dr. C. Dumas of the Department Reconnaissance Cellulaire et Amelioration des Plantes, Universitè Cl. Bernard-Lyon I for their stimulating discussions and encouragement. We also greatly appreciate the support from the Commission of the European Communities, through the ERASMUS programme.

We thank Professor H.G. Dickinson for providing the Foreword to this volume.

We are very grateful to the many authors who have made micrographs available for the atlas: Ms S.H. Barnes (London), Dr. F. Bouman (Amsterdam), Dr. M. de Boer-de Jeu (Wageningen), Mr. F. Ciampolini (Siena), Prof. H. Dickinson (Oxford), Prof.I. K. Ferguson (Kew), Prof.dr. P. Hepler (Amherst), Prof. J. Heslop-Harrison (Aberystwyth), Dr. Bing-Quan Huang (Norman), Dr. J. Janson (Wageningen), Prof. M. H. Kurmann (Kew), Dr. C. Keijzer (Wageningen), Mr. C. Milanesi (Siena), Dr. M. Murgia (Siena), Prof.dr. E. Pacini (Siena), Dr. E. Pierson (Siena), Dr. M. Rougier (Lyon), Dr. J. Schel (Wageningen), Dr. C. Theunis (Wageningen), Dr. A. Tiezzi (Siena), Mr. A. van Aelst (Wageningen), Dr. A. van Lammeren (Wageningen), and Dr. H. Wilms (Wageningen).

The facilities of the Department of Environmental Biology, University of Siena, and the Department of Botany, Agricultural University of Wageningen, have made it possible to prepare the "Atlas of Sexual Plant Reproduction". Our sincere appreciation is extended to Ms. C. Faleri for processing the photographs, Mr. L. Borghi and Mr. F. Vanni for preparing the plates, to Mr. A. Haasdijk and Mr. P. Snippenburg for preparing the drawings, and to Ms. T. van de Hoef-van Espelo, Ms. R. van den Brink-de Jong, Ms. A. Tanganelli and Ms. P. Barbi for typing the texts.

M. Cresti   S. Blackmore   J. van Went

CONTENTS

PART 3: PROGAMIC PHASE AND FERTILIZATION 163

Introduction 165

GENERAL INTRODUCTION

General Introduction

Sexual reproduction in flowering plants requires the coordinated development of the two reproductive organs of the flower, the anther and the pistil, and their successful interaction. The basic processes of sexual reproduction are meiosis and the fusion of gametes. The former results in the rearrangement of the genes and the reduction of the number of chromosomes and the latter results in the restoration of the original diploid chromosome number. During meiosis new gene combinations can be formed, and through gametic fusion new combinations of chromosomes can be established.

Both the anther and pistil show characteristic structures and developmental pathways. In the anther, the male reproductive organ, the sporogenous tissue is composed of microspore mother cells, which undergo meiosis to produce tetrads of haploid microspores. These cells divide once or twice during their development to become the male gametophytes or pollen grains. Where a single mitotic division takes place the mature pollen grain is composed of a vegetative cell and a generative cell. In plant species with bicellular pollen grains the generative cell divides to form two sperm cells after germination of the grain. This division occurs during pollen maturation in plants with tricellular pollen grains.

In the pistil, the female reproductive organ, a megaspore mother cell, which undergoes meiosis, is formed in each ovule. Of the four resulting megaspores, only one develops into a female gametophyte, the embryo sac. The mature embryo sac usually contains seven cells: the egg cell, two synergids, the central cell and three antipodal cells.

After maturation and dehiscence of the anther, the pollen grains are carried to the stigma, where they form pollen tubes which grow through the style towards the ovules. Ultimately, in this progamic phase, the generative cell divides into two sperm cells which become positioned at the tip of the pollen tube. The pollen tube enters the micropyle of the ovule and penetrates one of the synergids where it opens and releases the sperm cells. One sperm cell fuses with the egg cell to form the diploid zygote, whilst the other sperm cell fuses with the central cell to form the first cell of the triploid endosperm. Together the two fusions are known as double fertilization. Subsequently, the zygote begins rapid division to form the embryo, while the endosperm rapidly divides and expands to form the major tissue of the seed, rich in reserves.

Microscopical techniques have played a dominant role in the study of sexual reproduction of angiosperm plants. The structures involved all are very minute, and significant components of the process are not directly observable since they take place in gametophytic tissues that are deeply embedded in the surrounding tissues of the sporophyte. Microscopical research not only requires good microscopes, but appropriate techniques of fixation and preparation. The dramatic progress achieved in the study of plant sexual reproduction has closely mirrored the development of such facilities.

The development of transmission and scanning electron microscopes, together with appropriate new preparation techniques, have resulted in a wave of new information leading to the better understanding of the processes of sexual reproduction in angiosperms. The aim of this atlas is to present a broad survey of this knowledge and to demonstrate the ultrastructural aspects of the basic phenomena through selected illustrations.

The atlas is organized in three parts:
1. "Anther development" describes the development of the anther leading to the release of functional pollen grains.
2. "Pistil development" illustrates the development of the pistil up to the receptive stage.
3. "Progamic phase and fertilization" covers the progamic phase and the processes of fertilization which lead to the formation of the embryo and endosperm.

Recommended literature
Cresti M, Gori P, Pacini E (eds) (1988) Sexual reproduction in higher plants. Springer, Berlin-Heidelberg-New York

4

Johri B M (ed) (1984) Embryology of angiosperms. Springer, Berlin-Heidelberg-New York-Tokyo

Linskens H F (ed) (1974) Fertilization in higher plants. North Holland, Amsterdam

Linskens H F, Heslop-Harrison J (eds) (1984) Cellular interactions. Encycl Plant Physiol Vol 17. Springer, Berlin-Heidelberg-New York-Tokyo

Shivanna K R, Johri B M (1984) The angiosperm pollen: structure and function. Wiley Eastern, New Delhi

List of abbreviations

A	antipodal cell
AW	anther wall
Ba	baculum
Ca	callose
CaP	callosic plug
CC	central cell
Chr	chromosome
Co	coleoptile
CP	compatible pollen grain
CW	cell wall
D	dictyosome (Golgi body)
dS	degenerated synergid
E	exine
EC	egg cell
ECT	ectexine
Em	embryo
En	endothecium
END	endexine
EN	egg cell nucleus
End	endosperm
ER	endoplasmic reticulum
ES	embryo sac
Ex	exothecium
Exu	exudate
F	funiculus
FA	filiform apparatus
GC	generative cell
GCW	generative cell wall
GN	generative nucleus
GP	germination pore
Gv	Golgi-vesicle
GZ	glandular zone
H	hypostase

I	intine
II	inner integument
IP	incompatible pollen grain
IS	intercellular substance
IT	integumentary tapetum
L	lipid body
M	mitochondrium
MC	megaspore mother cell
Mi	micropyle
ML	middle layer
MMC	microspore mother cell
MS	microspore
Mt	microtubules
N	nucleus
Nu	nucleolus
NUC	nucellus
OI	outer integument
P	plastid
Pa	parenchyma
Pe	pellicle
PG	pollen grain
Pl	placenta
PM	plasma membrane
PN	polar nucleus
Pr	primexine
PS	persistent synergid
PT	pollen tube
R	ribosomes
RER	rough endoplasmic reticulum
S	starch
Sc	scutellum
SC	sperm cell
SCN	sperm cell nucleus
SER	smooth endoplasmic reticulum
Sm	shoot meristem
SP	stigma papilla
ST	sporogenous tissue
StC	stylar canal
Sus	suspensor
Sy	synergid

T	tapetum
Ta	tannin
Te	tectum
TT	transmitting tissue
U	Ubisch bodies
V	vacuole
VB	vascular bundle
VC	vegetative cell
VN	vegetative nucleus
Z	zygote

PART 1: ANTHER DEVELOPMENT

ANTHER DEVELOPMENT

Introduction

In most angiosperms each stamen is composed of an anther and a filament. The anther usually contains four microsporangia and an intervening connective which is linked with the filament. The anther wall consists of four layers: the epidermis (exothecium), endothecium, middle layer(s) and tapetum. The central region of each microsporangium contains the sporogenous tissue, composed of microspore mother cells (meiocytes), which eventually will form the pollen grains.

Two phases can be distinguished in the development of the sporogenous tissue, microsporogenesis and microgametogenesis. Microsporogenesis comprises the (meiotic) reduction division and the formation of individual microspores whilst microgametogenesis encompasses the subsequent development of microspores into mature bicellular or tricellular pollen grains, which contain the gametes.

At the onset of meiosis the microspore mother cells (meiocytes) are large, rich in cytoplasm, and have only thin pecto-cellulosic walls. During early prophase the meiocytes become interconnected by cytoplasmic channels and a dedifferentiation of the cytoplasm commences. This is especially evident in a sharp decrease in the number of ribosomes present in the cytoplasm and in structural changes of the organelles, which generally become much simpler in organisation.
Dedifferentiation is thought to be related to a transition from sporophytic to gametophytic gene expression accompanying the change from diploid to haploid generation. As prophase commences, deposition of callose begins along the plasma membranes, resulting in thick callosic walls separating the meiocytes from each other.

Meiosis results in the formation of four haploid nuclei, which become separated through cytokinesis, accompanied by cell plate formation. The timing of cytokinesis varies and may be either successive or simultaneous. In successive cytokinesis, cell plates are formed after both the first and second meiotic division so that there is a distinct dyad stage, whereas in simultaneous cytokinesis they are formed only after the second division. During cell plate formation additional deposition of callose takes place, resulting in thick callosic cell walls that separate the microspores from each other.

During microsporogenesis distinct changes occur in the anther wall. The middle layer(s) usually degenerate gradually and ultimately disappear. The cells of the tapetum, in contrast, enlarge and develop a complex ultrastructure, which indicates that they become very active metabolically. Up to this stage the exothecium and endothecium remain relatively unaltered, although the cells can enlarge and become more vacuolate.

The microspores start to differentiate whilst still associated in tetrads and encapsulated by callosic walls. Differentiation is accompanied by the restoration of the ribosome population. Nucleoloids which are frequently observed in the cytoplasm during microspore differentiation are regarded as stored ribosomes. Changes in organelle number and structure accompany the initiation of microspore wall formation which starts with the establishment of the cellulosic primexine which forms a template on which sporopollenin is subsequently deposited. The first sporopollenin to be deposited is synthesized in the young microspores, but subsequently sporopollenin is also produced by the tapetal cells. The final pattern and morphology of the outer pollen wall, the exine, is usually already determined, even at this early stage.

After the establishment of a first layer of sporopollenin forming the young exine, degradation of the callosic layer starts as callase is produced by the tapetal cells. Exine deposition continues after release of the microspores from the tetrad and finally an inner pecto-cellulosic cell wall, the intine, is layed down. After their release, the young microspores enlarge, which is accompanied by strong vacuolation. At the vacuolate microspore stage, the microspore nucleus undergoes the first mitotic division, followed by cytokinesis. As a rule cytokinesis results in a large vegetative cell and a small, lenticularly shaped generative cell which is attached to the intine. Frequently, during cytokinesis the organelles of the microspore are positioned in such a way, that most of them are transferred directly to the vegetative cell. This holds especially for the plastids, with the result that often the generative cells are completely lacking plastids.

At the time of the mitotic division of the microspore nucleus, the nuclei of the tapetal cells also divide. Here however, mitosis is not followed by cytokinesis and a binucleate or polyploid condition arises in the tapetal cells. The subsequent development of the tapetum shows great diversity and two major tapetal types are usually distinguished. The glandular or secretory type is characterised by cells which remain intact and persist in situ, whereas in the amoeboid or periplasmodial type the cell walls break down and the protoplasts intrude into the locule, eventually forming a coenocytic plasmodium.

During microgametogenesis and maturation of the pollen grain the vegetative cell shows a number of marked cytoplasmic changes, related to its future function as carrier of the male gametes. The cell becomes rich in cytoplasm, with only small vacuoles, and vast amounts of reserves are formed, consisting of various combinations of starch, lipids and proteins. The final composition of the organelle population in the mature vegetative cell shows great diversity among species. In many, large quantities of rough endoplasmic reticulum with the cisternae arranged in stacks are found. In other species, the pollen grains contain large numbers of dictyosomes, or abundant dictyosome-vesicles. In spite of this great diversity, the vegetative cell of the pollen grain can always be regarded as a storage cell, equipped for the future formation of the pollen tube, and transmission of the male gametes.

Concurrently with the development of the vegetative cell, the generative cell also undergoes a number of changes. After its formation, the generative cell separates from the intine and moves to a position where it is completely enclosed by the vegetative cell. During this relocation, its shape changes, becoming first spherical and then spindle-shaped. The establishment of the spindle shape is accompanied by the formation of numerous bundles of microtubules, which become located in the peripheral cytoplasmic region and positioned parallel to the long axis of the cell. Whereas the newly formed generative cell has a normal cell wall, the spindle-shaped generative cell is surrounded simply by a thin layer separating the plasma membranes of the two cells. In many species, the generative cell divides into two spindle-shaped sperm cells, before germination of the pollen tube. The two sperm cells frequently remain connected to each other and become located near the vegetative nucleus, forming the so-called male germ unit. Both generative cells and sperm cells show a relatively simple ultrastructure with only few organelles.

In synchronization with the developing pollen grains, two tissues of the anther also differentiate to complete their functions. The cells of the tapetum, whether secretory or amoeboid, contain numerous organelles among which rough and smooth endoplasmic reticulum are especially prominent, indicating the active synthesis of many substances, needed for the development of functional pollen. Of particular importance are the formation and the deposition on the pollen surface of sporophytic proteins and Pollenkitt. Before the anther matures, the tapetum degenerates and its remains become deposited on the pollen as the tryphine.

Anther maturation is accompanied by considerable loss of water and dehydration of the pollen grains. Just before maturation, the endothecial cells acquire fibrous thickenings of their radial cell walls. These thickenings cause tangential shrinkage during anther dehydration, leading to the rupture of the anther wall and the release of the pollen grains.

The mature pollen grains are well protected and fully equipped to complete their function as carriers of the male gametes . The mature pollen wall is composed of an outer sporopollenin layer, the exine, and an inner polysaccharide layer, the intine. The internal structure and surface morphology of the exine is highly variable and often taxon-specific. The form is related to the mode of pollen dispersal, among other factors. In wind pollination, the thickness, surface patterning and stickiness of the exine are generally reduced. In biotically pollinated plants the exine is generally elaborate and the pollen grains are usually very sticky, enabling them to stick to each other, to the vector and to the stigmatic surface.

Recommended literature

Blackmore S, Knox R B (eds) (1990) Microspores: evolution and ontogeny. Academic Press, London-San Diego-New York-Tokyo

Cresti M, Gori P, Pacini E (eds) (1988) Sexual reproduction in higher plants. Springer, Berlin-Heidelberg-New York

Heslop-Harrison J (ed) (1971) Pollen: development and physiology. Butterworths, London

Mulcahy D L, Bergamini Mulcahy G, Ottaviano E (eds) (1986) Biotechnology and ecology of pollen. Springer, New York-Berlin-Heidelberg-Tokyo

Stanley R G, Linskens H F (1974) Pollen biology, biochemistry, management. Springer, Berlin-Heidelberg-New York

Willemse M T M, Van Went J L (eds) (1985) Sexual reproduction in seed plants, ferns and mosses. Pudoc, Wageningen

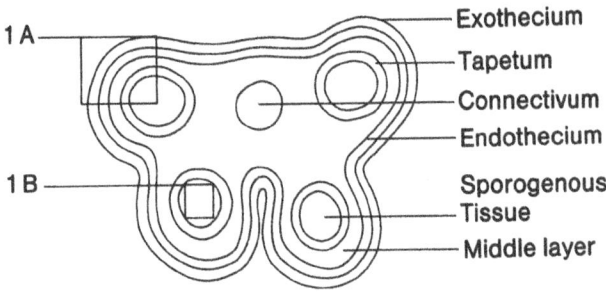

Plate 1A. Transverse section through an anther of *Brassica napus* L., showing the various tissues which compose the microsporangium: the exothecium, the endothecium, the middle layers, the tapetum, and the sporogenous tissue. Even at this early stage of development each tissue shows a specific cell shape and organization. x 3,000.

Plate 1B. Portion of the sporogenous tissue of *Brassica oleracea* L., at late prophase of microsporogenesis. The cells are very cytoplasmic rich, and have very thin pecto-cellulosic cell walls. Intercellular spaces are absent. x 4,200.

(Plate 1A courtesy M Murgia, Siena; Plate 1B courtesy M Rougier, Lyon).

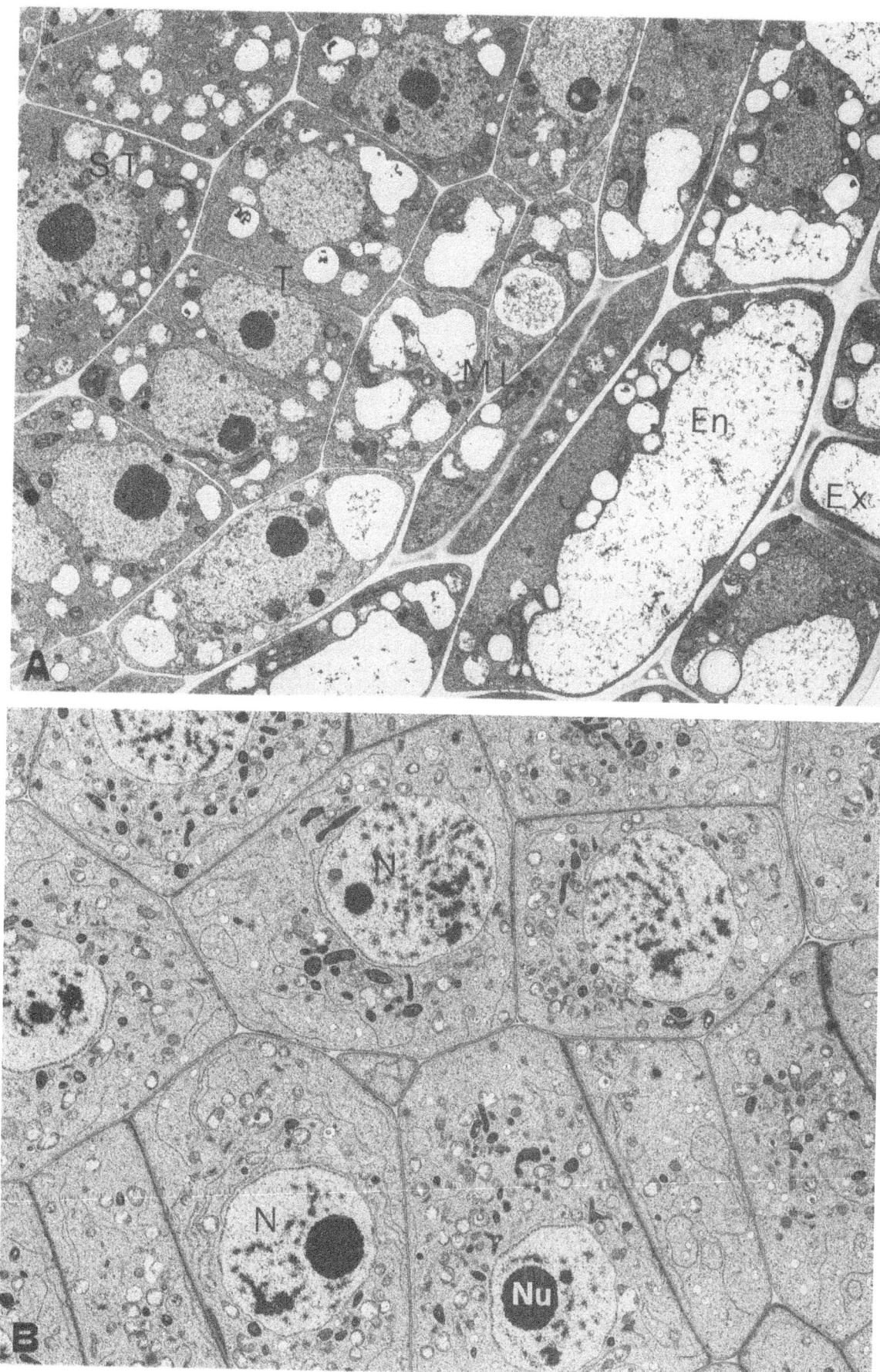

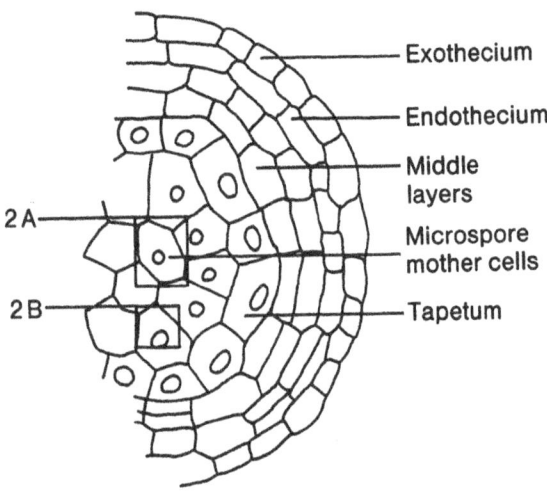

Exothecium

Endothecium

Middle layers

Microspore mother cells

Tapetum

Plate 2A. Microspore mother cell of *Brassica oleracea* L., at late prophase. The cytoplasm has a relatively simple ultrastructure, and most constituents appear to be randomly distributed. In the nucleus chromosomes are formed, and the process of chromosome pairing is initiated. The positioning of the nuclear pores at one specific portion of the nuclear envelope is remarkable (arrows). x 10,000.

Plate 2B. Ultrastructural details of a microspore mother cell of *Impatiens sultani* Hook f. at late prophase stage. The cytoplasm has a very "diluted" appearance. The arrow indicates a nucleoloid, an aggregation of ribosomes. In the nuclear envelope many pores are present (arrowheads). x 15,000.

(Plate 2A courtesy M Rougier, Lyon; Plate 2B reproduced by permission from: Van Went J, Cresti M (1989) Protoplasma 148: 1-7).

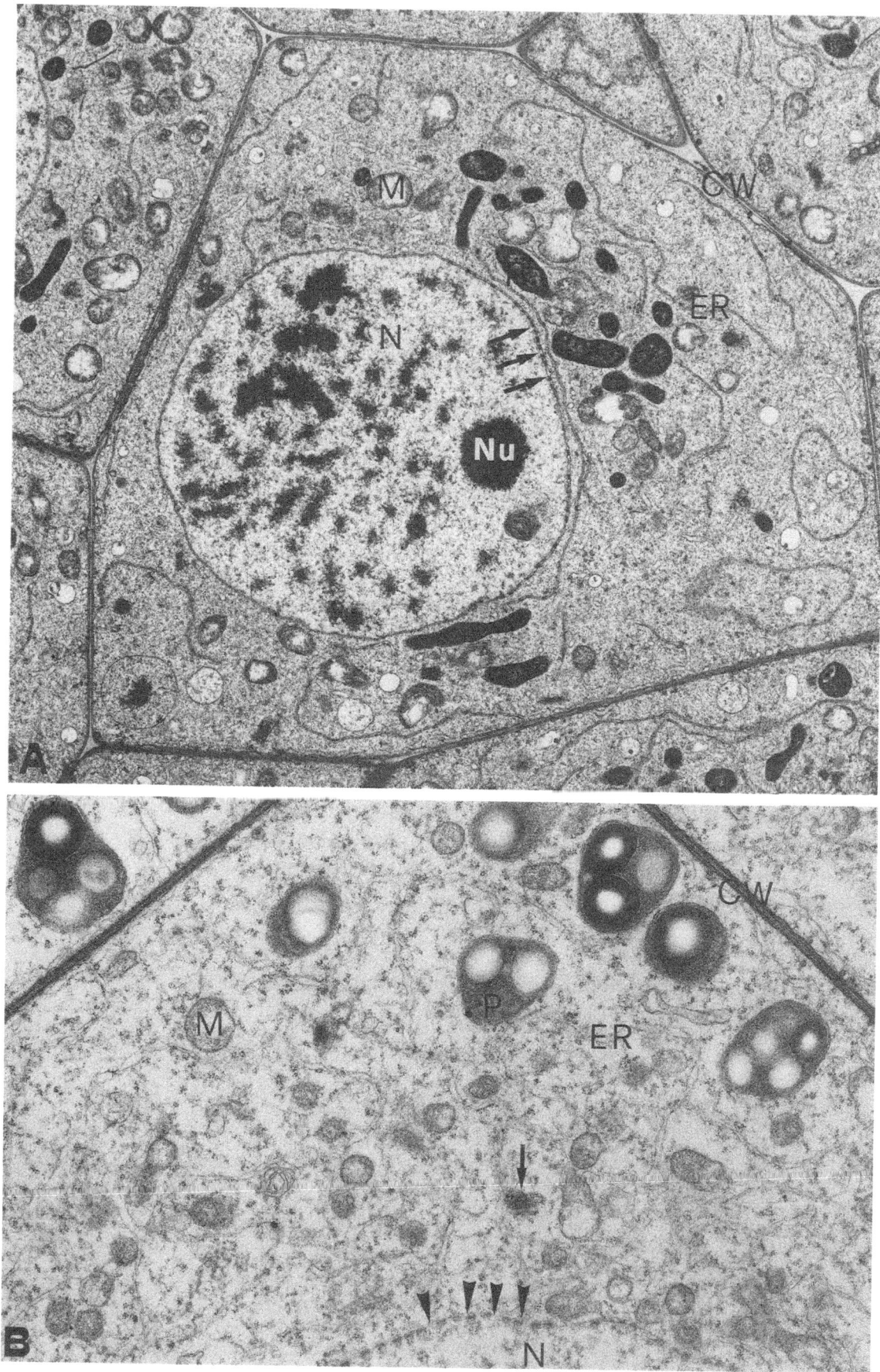

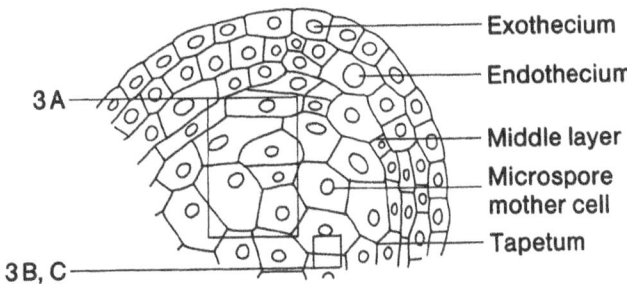

Plate 3A. Sporogenous tissue and surrounding tapetum of *Catananche caerulea* L., at early stage of microsporogenesis. The anthers have been chemically fixed, freeze-fractured, and critical point dried to prepare them for scanning electron microscopy. x 2,000.

Plate 3B. Enlarged portion of sporogenous cells of *Catananche caerulea* L., at early prophase. The cytoplasm of the cell is interconnected by channels (arrow). x 15,000.

Plate 3C. Portion of microspore mother cells of *Impatiens sultani* Hook f. at early stage of microsporogenesis. The cells are interconnected by cytoplasmic channels (arrowhead), which are thought to function in the synchronization of the development of the individual cells. The cytoplasm shows a relatively simple ultrastructure, marked by a low number of ribosomes. x 16,000.

(Plates 3A, B reproduced by permission from Blackmore S, Barnes S H (1988) Ann Bot 62: 605-614).

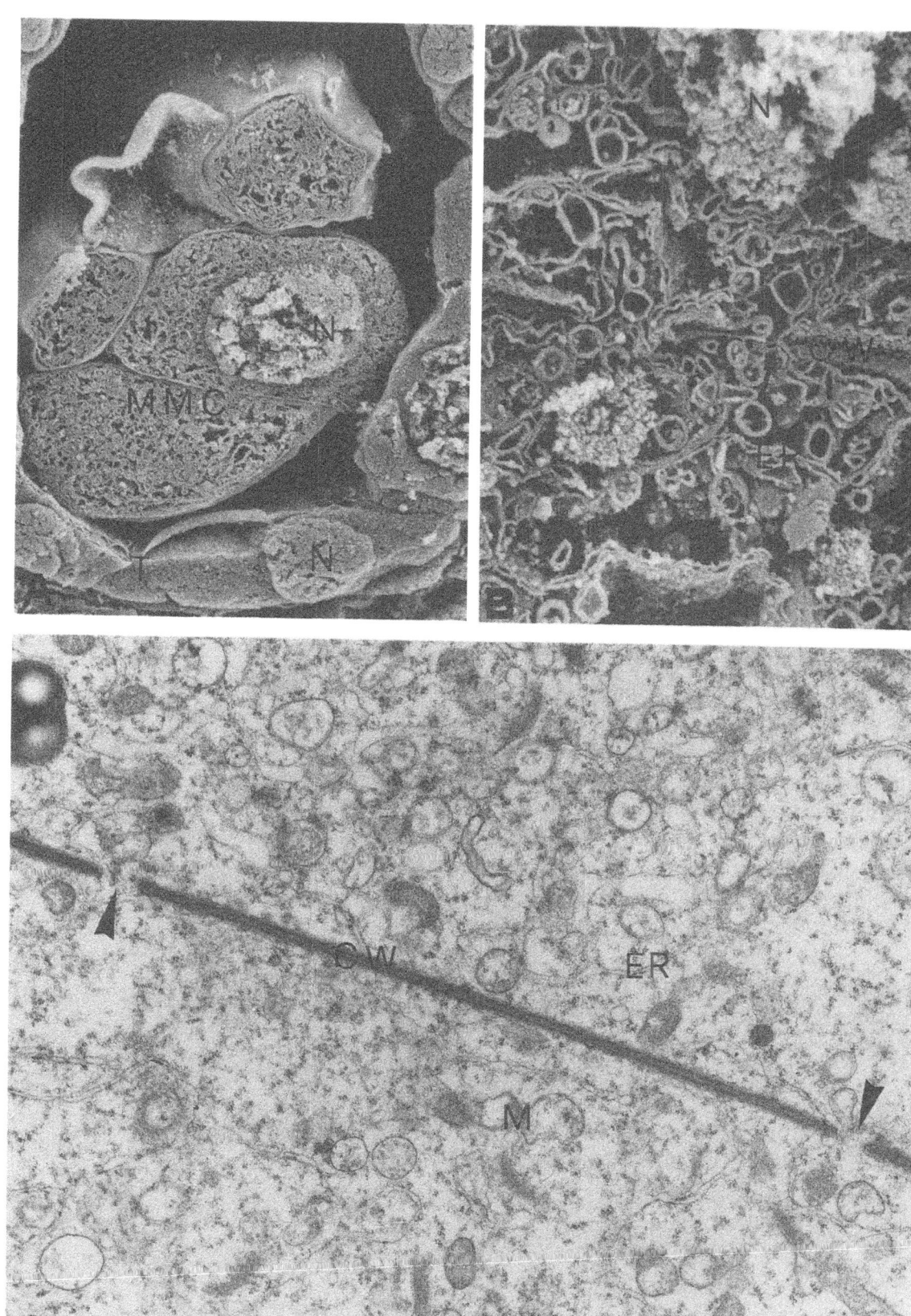

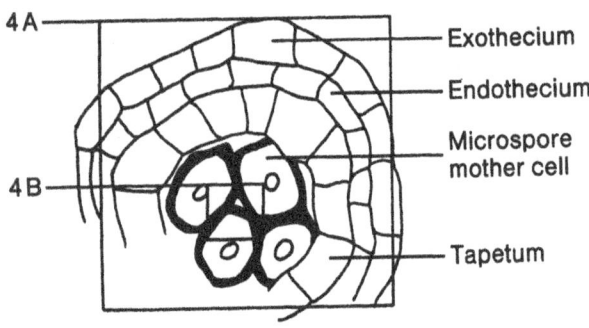

Plate 4A. Cross section through an anther of *Catananche caerulea* L. at mid-prophase of microsporogenesis. The anther has been chemically fixed, freeze-fractured and critical point-dried to prepare it for scanning electron microscopy. During the prophase callose is deposited around the microspore mother cells. Eventually a two-layered wall is present, composed of an outer, thin, pecto-cellulosic cell wall (original microspore mother cell wall), and an inner, thick callosic cell wall (arrows). During callosic wall formation most of the cytoplasmic connections between the microspore mother cells become interrupted, resulting in the further individualization of the cells. Simultaneously, the endoplasmic reticulum of the microspore mother cells assumes a characteristic arrangement of concentric sheets. By this stage the tapetal cells have intact cell walls. The cells have enlarged and many nuclei have undergone mitosis, leading to a binucleate condition of most of the cells. x 2,000.

Plate 4B. Enlarged portion of microspore mother cells of *Brassica napus* L., showing the formation of the callosic wall layer (arrows). x 19,000.

(Plate 4A reproduced by permission from Blackmore S, Barnes S H (1988) Ann Bot 62: 605-614; Plate 4B courtesy M Murgia, Siena).

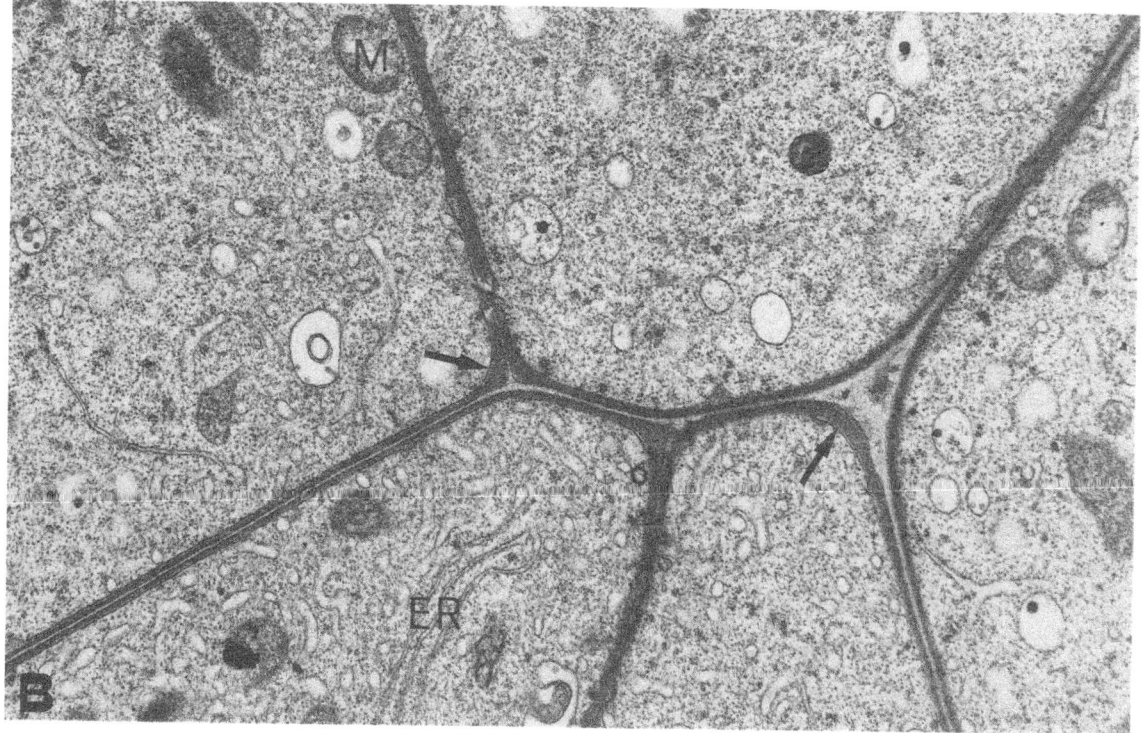

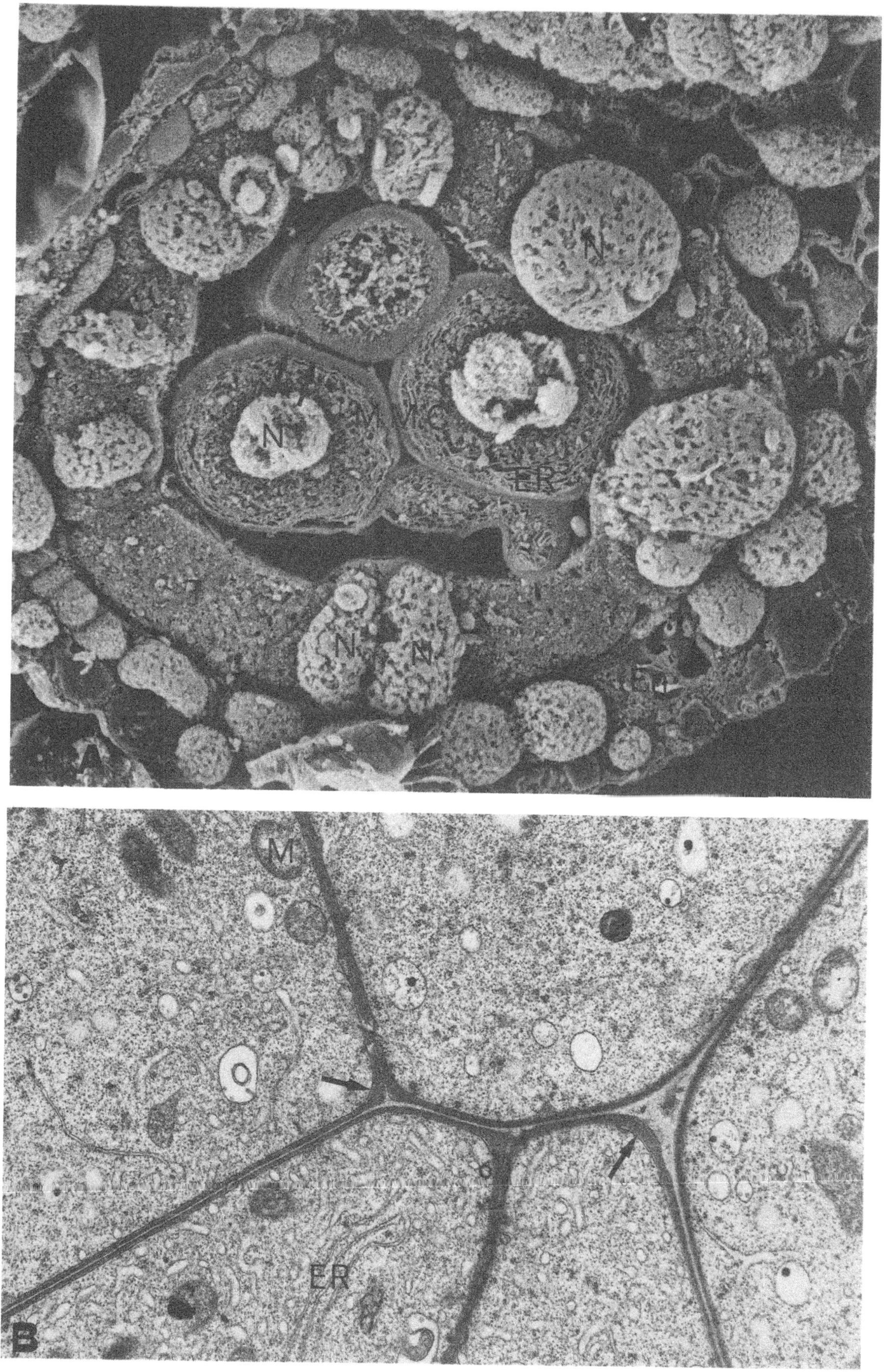

25

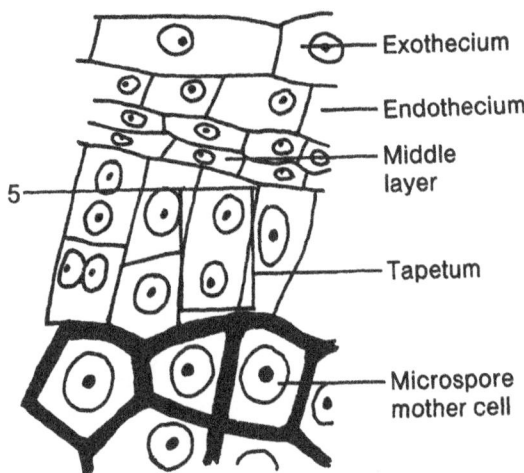

Exothecium

Endothecium

Middle layer

Tapetum

Microspore mother cell

Plate 5. Enlarged portion of a binucleate tapetum cell of *Brassica oleracea* L., showing the two nuclei and the complexly structured cytoplasm. The cytoplasm contains numerous mitochondria, plastids and free ribosomes. There is an extensive endoplasmic reticulum and dictyosomes are producing large Golgi-vesicles (arrow). x 17,000.

(Plate 5 courtesy M Murgia, Siena)

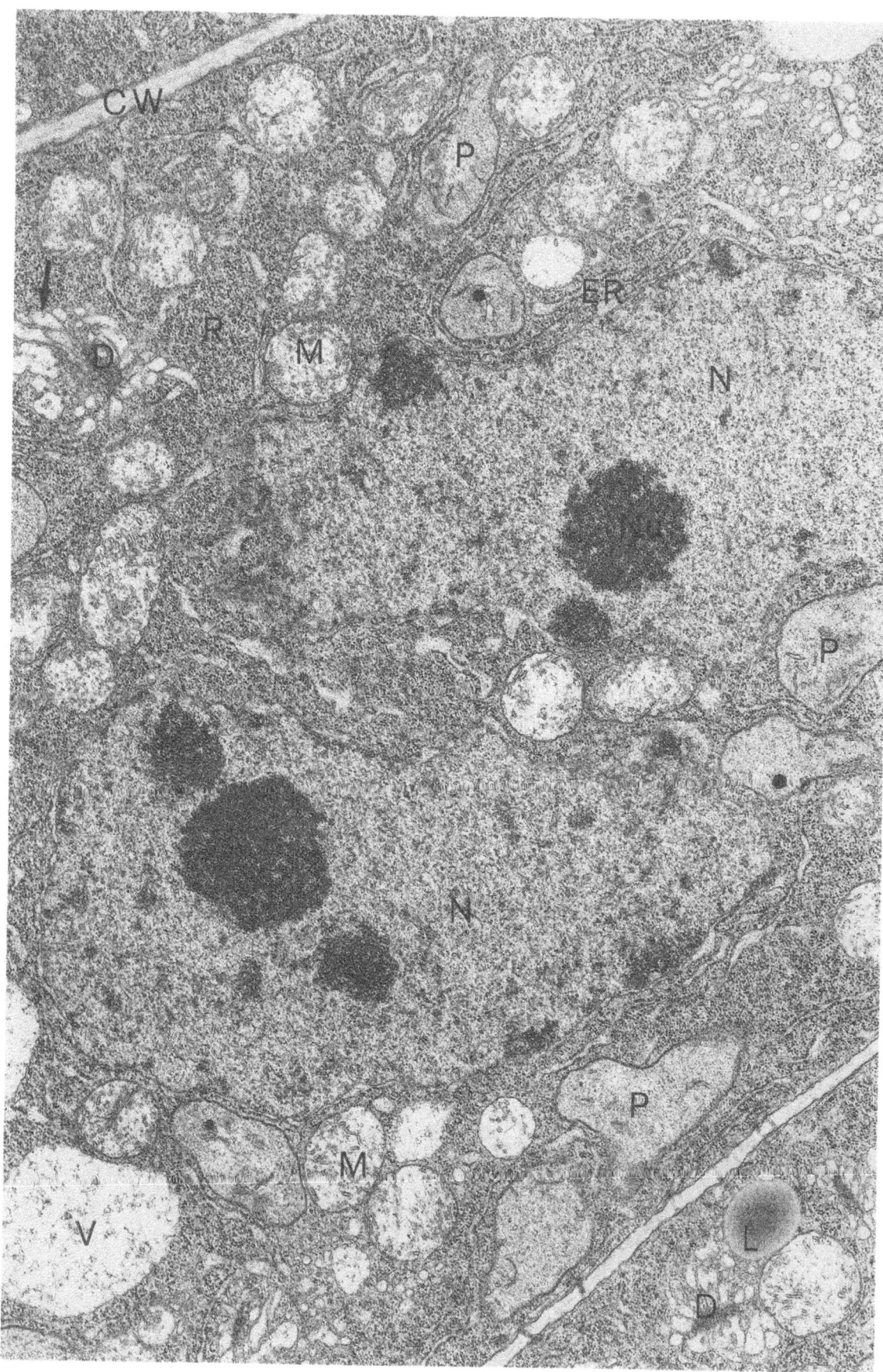

27

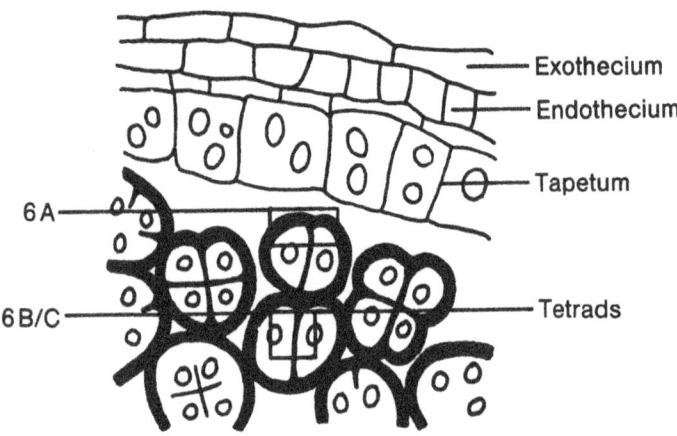

Plate 6A. Enlarged portion of a microspore mother cell of *Impatiens sultani* Hook f. at the onset of tetrad formation. Cytokinesis starts with the formation of local ingrowths (asterisk) of the callosic wall. Note the difference in appearance of the plasma membrane along the wall ingrowth (wrinkled) and adjacent wall parts (straight). x 17,000.

Plate 6B. Enlarged portion of a microspore mother cell of *Impatiens sultani* Hook f. at the onset of tetrad formation. Simultaneously with the local ingrowth of the surrounding callosic wall, cell plate formation starts with the accumulation of small Golgi-vesicles in the central region of the cell. x 17,000.

Plate 6C. Enlarged portion of a microspore mother cell of *Impatiens sultani* Hook f. during tetrad formation. The accumulated Golgi-vesicles fuse to form aggregates of tubules and small sacs (arrow). Subsequently deposition of callose takes place inside the sacs, leading to the formation of callosic cell plates (double arrow). x 35,000.

(Plates 6A, C reproduced by permission from Van Went J, Cresti M (1988b) Sex Plant Reprod 1: 228-233; Plate 6B reproduced by permission from Van Went J, Cresti M (1989) Protoplasma 148: 1-7).

28

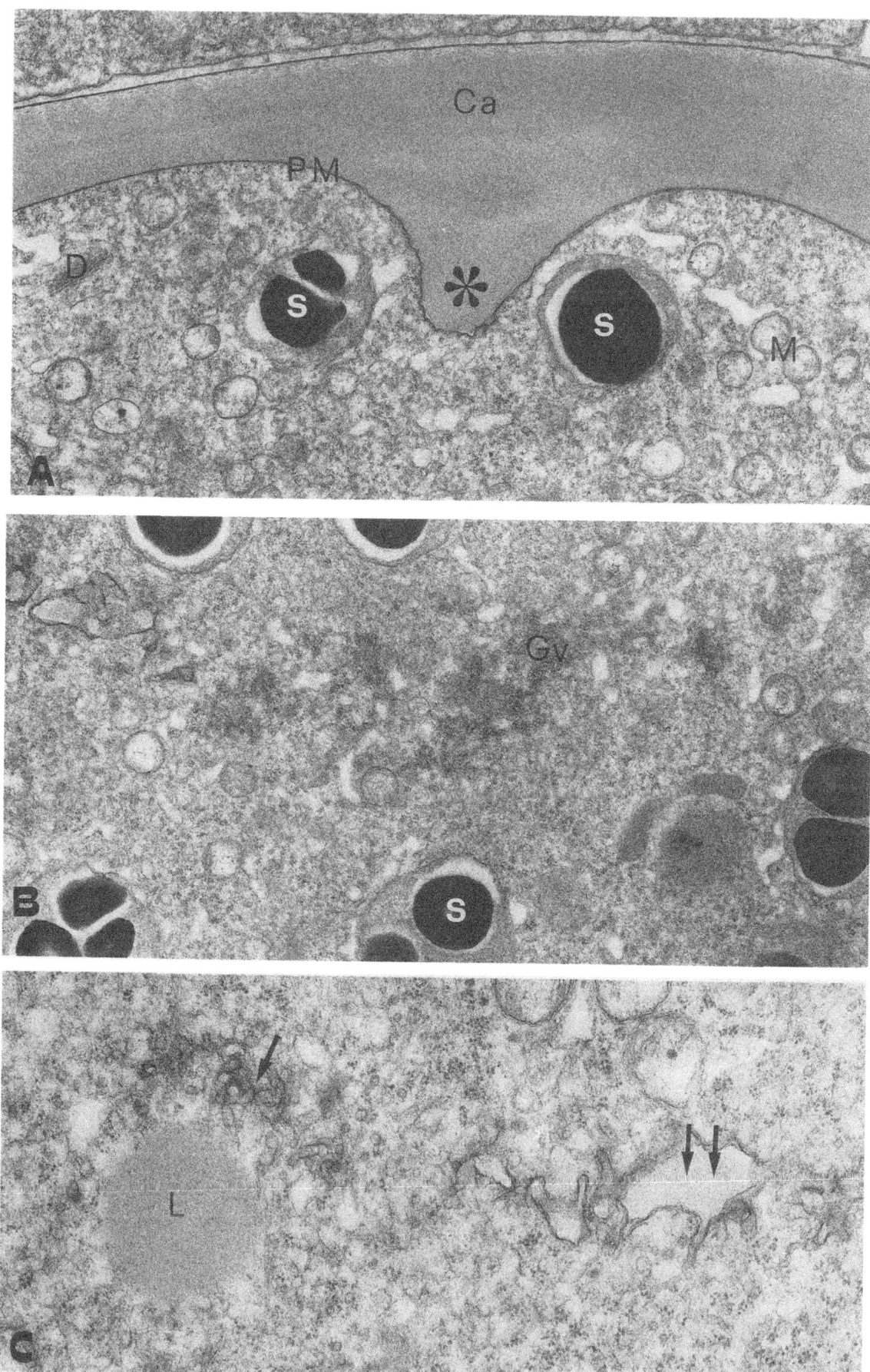

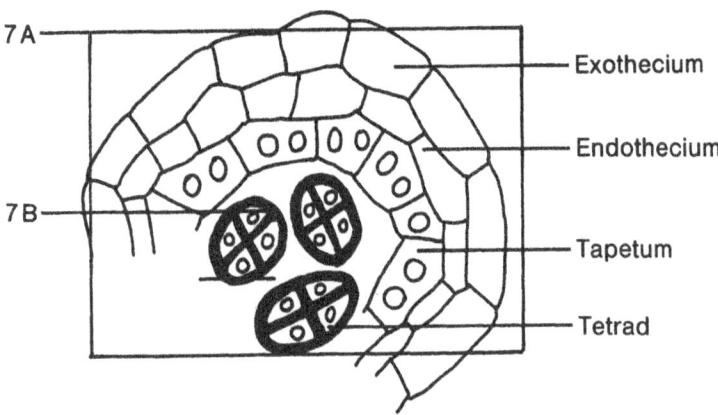

Exothecium

Endothecium

Tapetum

Tetrad

Plate 7A. Cross sectioned anther of *Catananche caerulea* L. at tetrad formation stage. The anther has been chemically fixed, freeze-fractured and critical point dried to prepare it for scanning electron microscopy. Following cell plate formation, additional callose deposition occurs. It starts from the wall ingrowths (arrows) and from the midpoint between the four tetrahedrally arranged microspores. x 2,000.

Plate 7B. Portion of a microspore mother cell of *Impatiens sultani* Hook f. at the final stage of cytokinesis. The initial wall ingrowths have enlarged centripetally. One of them has already fused with the callose deposit at the midpoint of the cell (asterisk). Note that the remaining cell plates connecting the other growing wall parts are still incomplete (arrows). x 7,100.

(Plate 7A reproduced by permission from Blackmore S, Barnes S H (1988) Ann Bot 62: 605-614; Plate 7B reproduced by permission from Van Went J, Cresti M (1988b) Sex Plant Reprod 1: 228-233).

Ca
T
N
N
T
Ca
N
En
Ex
A

Ca
a
P
Ca
*
N
N
B

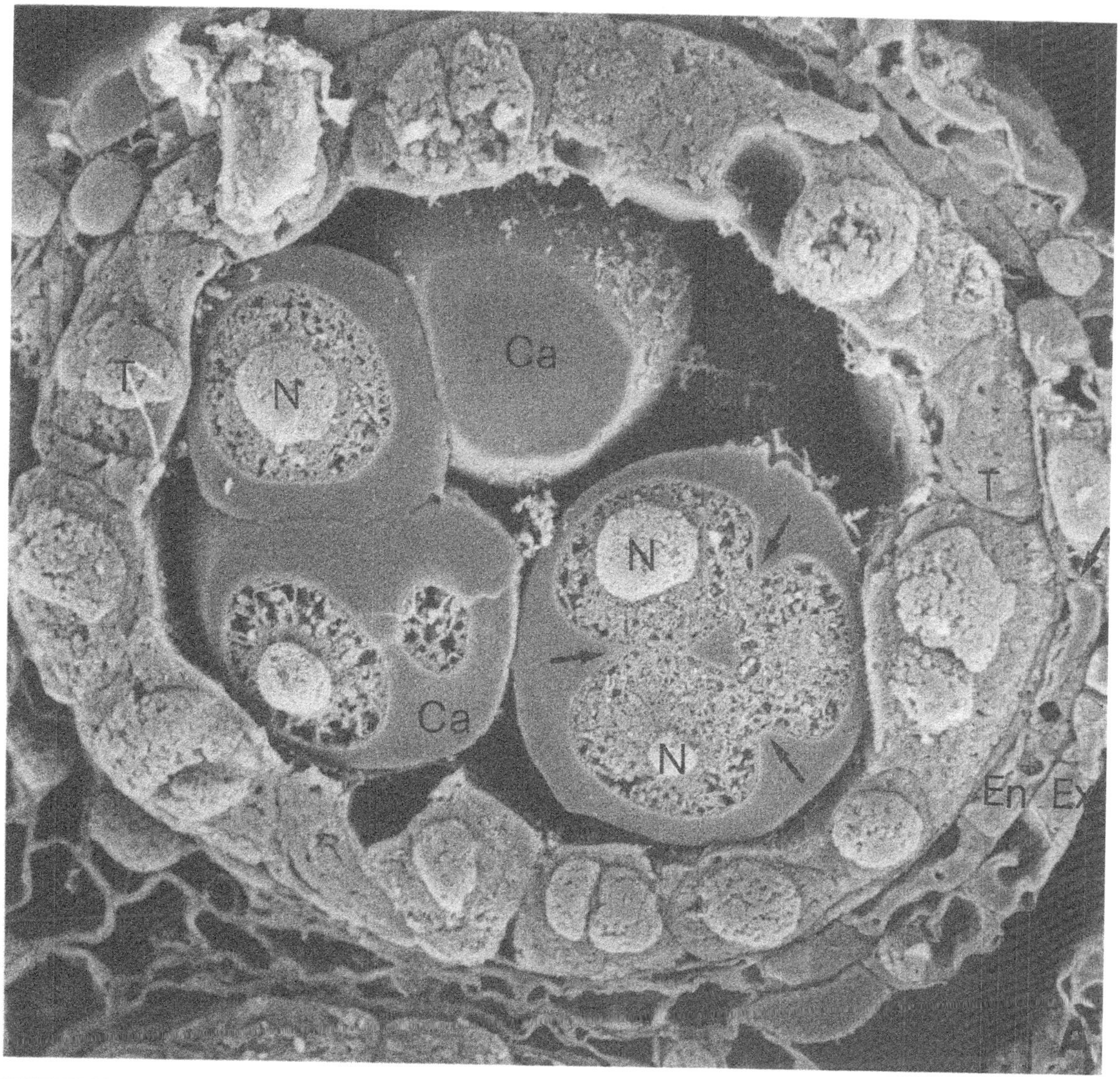

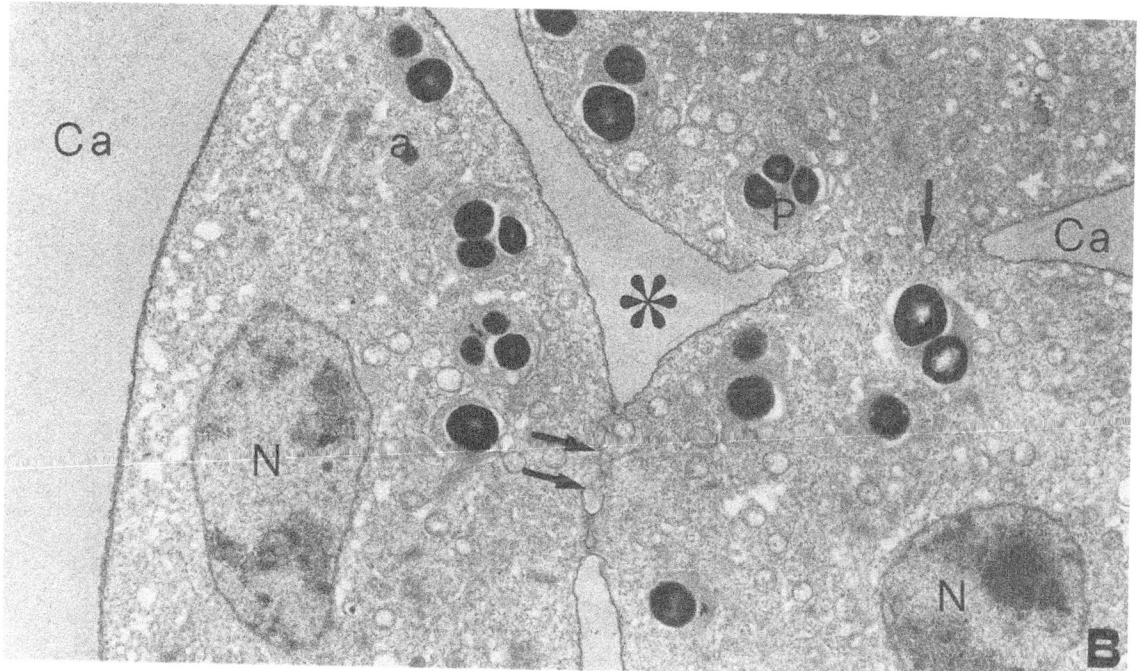

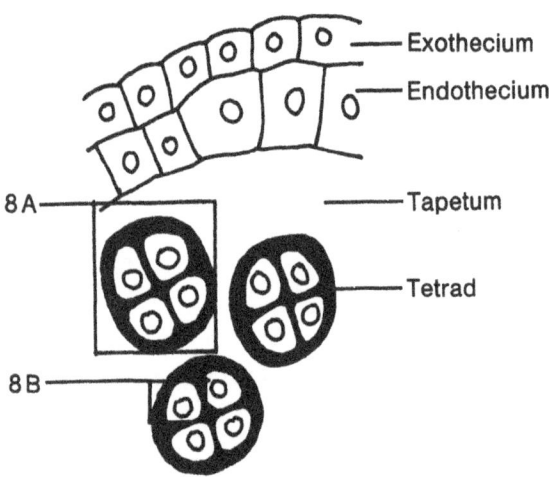

Plate 8A. Microspore mother cell of *Aloë ciliaris* Haw. after completion of cytokinesis. Cytokinesis results in the complete isolation of the four products of meiosis, the microspores. The microspores are not interconnected by plasmodesmata, and each of them develops into an individual gametophyte, the pollen grain. x 5,300.

Plate 8B. Enlarged portion of a microspore of *Impatiens sultani* Hook f., at the onset of primexine formation. While still within the tetrad, encapseled by callose, the development of microspore to pollen grain starts with the formation of a first gametophytic cell wall, called the primexine. Primexine formation involves the production and secretion of cell wall precursors by the dictyosomes and their vesicles. At this stage, nucleoloids are also present in the cytoplasm (arrow). Nucleoloids are interpreted as an indication for ribosome repopulation, associated with gametophytic gene expression. x 27,000.

(Plate 8B reproduced by permission from Van Went J, Cresti M (1989) Protoplasma 148: 1-7).

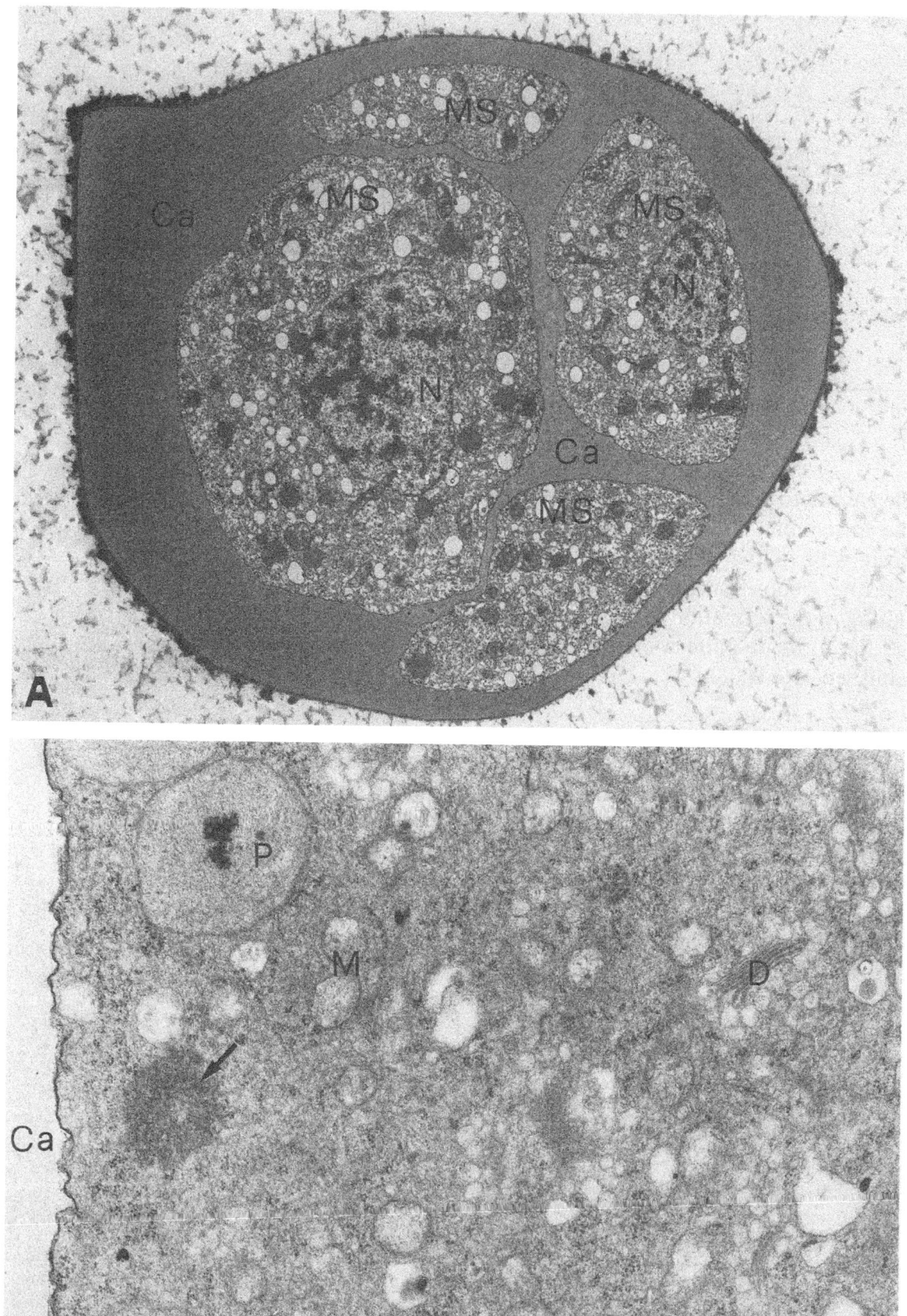

33

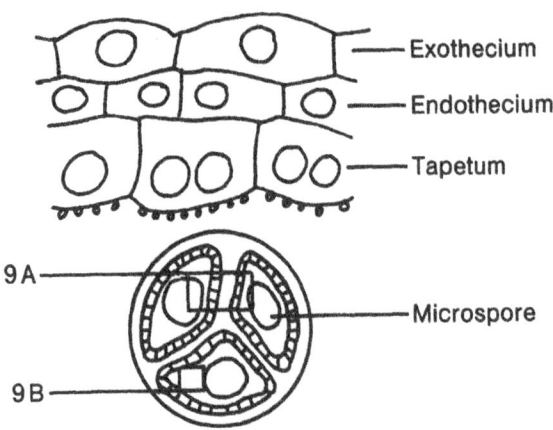

Exothecium

Endothecium

Tapetum

Microspore

Plate 9A. Microspores of *Impatiens sultani* Hook f. at early stage of exine formation. The microspores are still enclosed within the callosic wall of the tetrad (asterisk). After completion of the primexine, deposition of sporopollenin starts at specific sites, resulting in the formation of probaculae (arrows). Already at this stage the ultimate pattern of sporopollenin deposition becomes established. x 18,000.

Plate 9B.Enlarged portion of a microspore of *Impatiens sultani* Hook f. at probaculum formation stage. The transition from primexine to probaculae formation is accompanied by ultrastructural changes in the cytoplasm (compare with plate 8B). The production of Golgi-vesicles apparently has ceased, and instead an extensive smooth endoplasmic reticulum is formed. The ribosomes become accumulated in restricted areas of the cytoplasm (asterisk). x 35,000.

(Plates 9A, B reproduced by permission from Van Went J, Cresti M (1989) Protoplasma 148: 1-7).

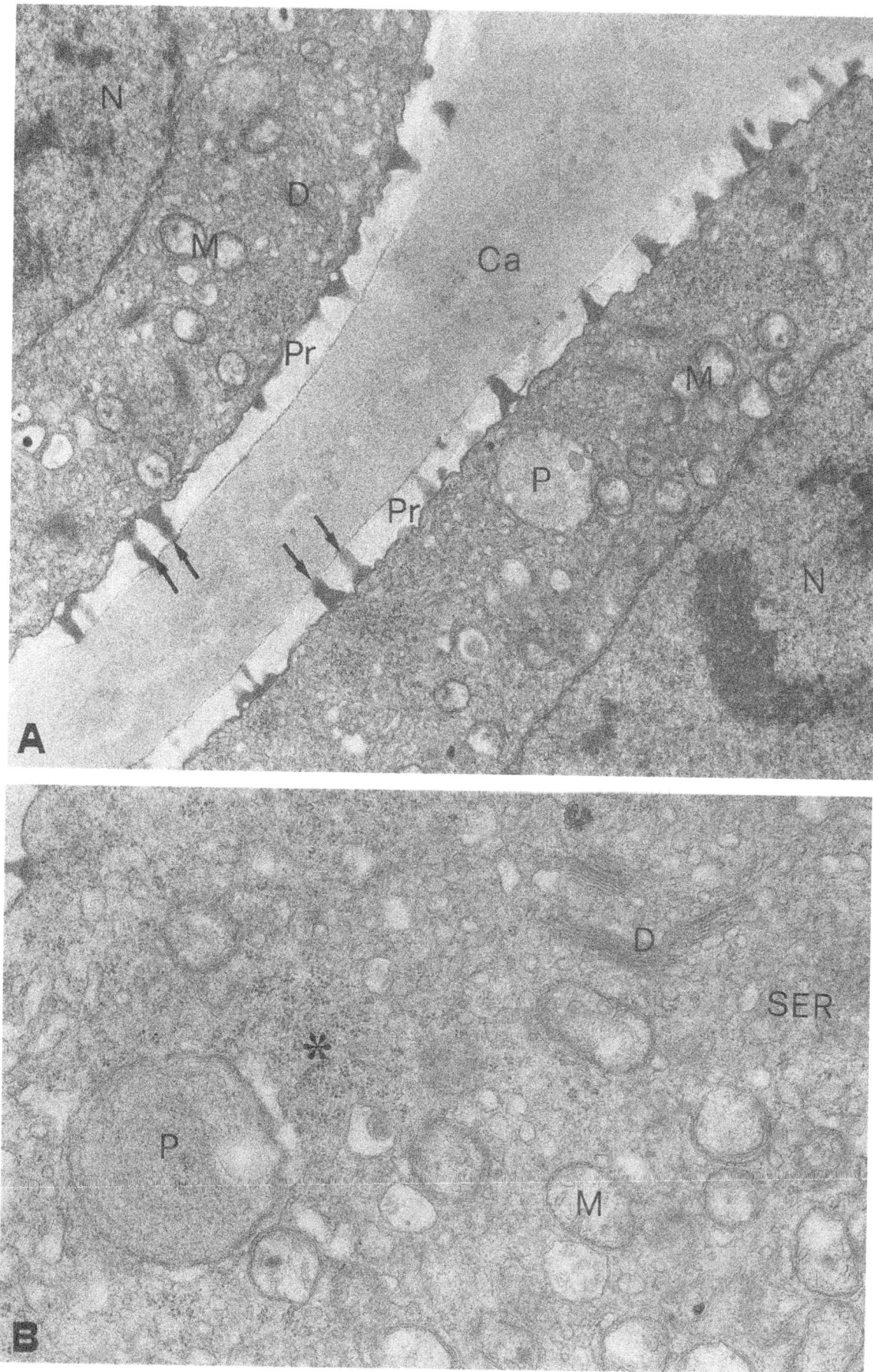

35

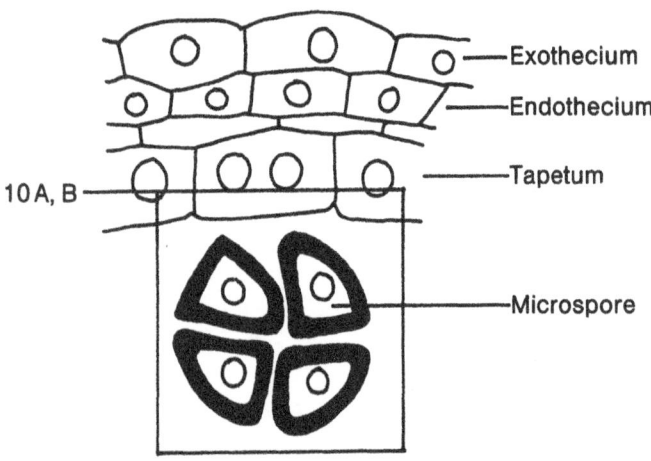

Plate 10A. Tetrad of *Brassica napus* L. during the dissolution of the tetrad callosic wall and the liberation of the microspores. When the sporopollenine wall of the microspore, the exine, has been formed to a certain extend, the surrounding callosic layer begins to dissolve. Dissolution of the callose starts from the outside, progressing gradually into the intercellular portions of the layer. This indicates that the enzymes involved in callose breakdown are produced by the tapetal cells and secreted into the anther locular space. x 3,000.

Plate 10B. Free microspores and adjacent tapetal cells of *Impatiens walleriana* Hook f., just after dissolution of the tetrad callose. At this stage the microspores are still densely packed with cytoplasm and cell enlargement has not yet been started. The binucleate tapetal cells loose their cell walls and become amoeboid. x 3,000.

(Plate 10A courtesy M Rougier, Lyon; Plate 10B reproduced by permission from Van Went J (1981) Acta Soc Bot Pol 50: 249-252).

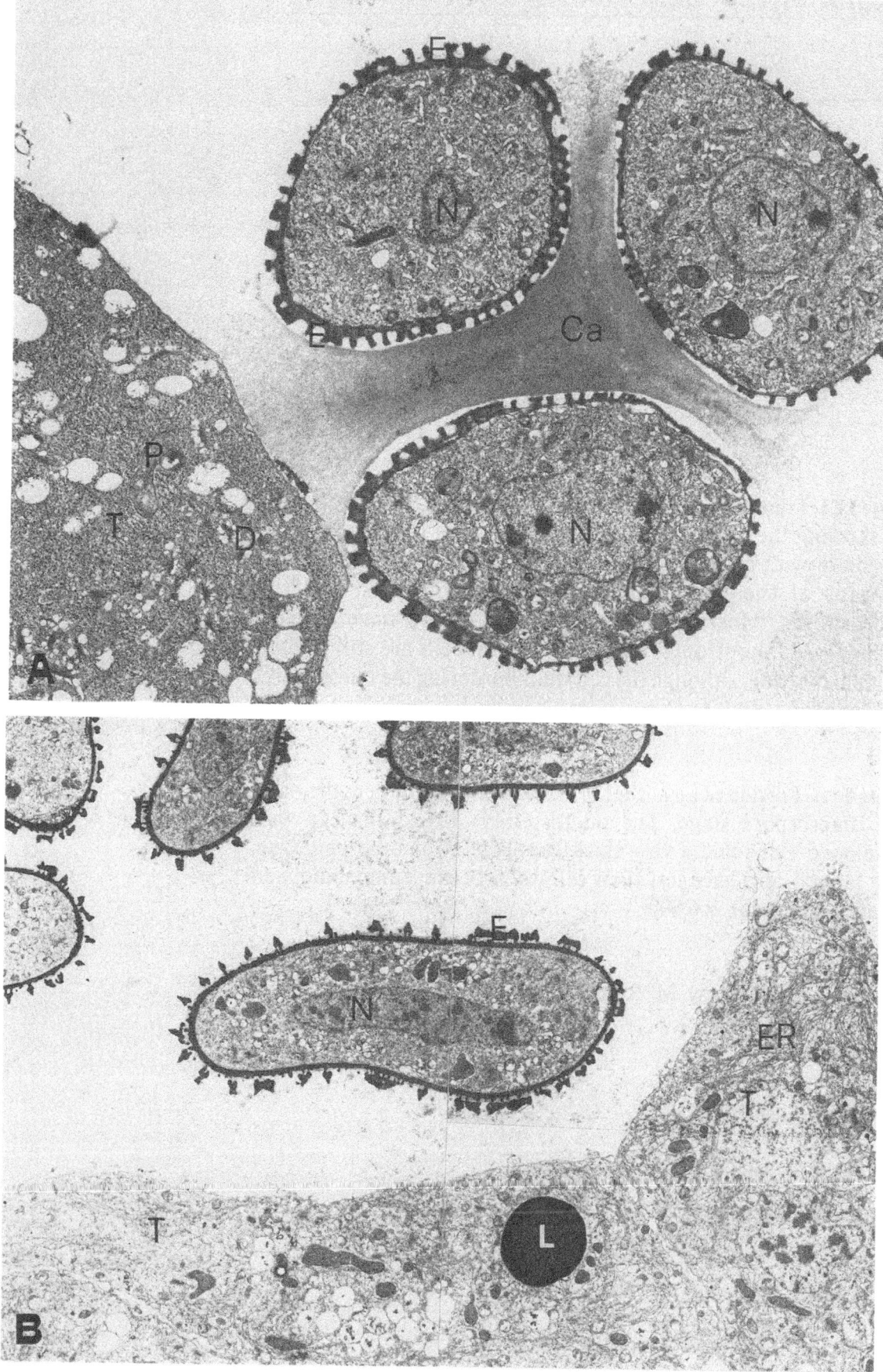

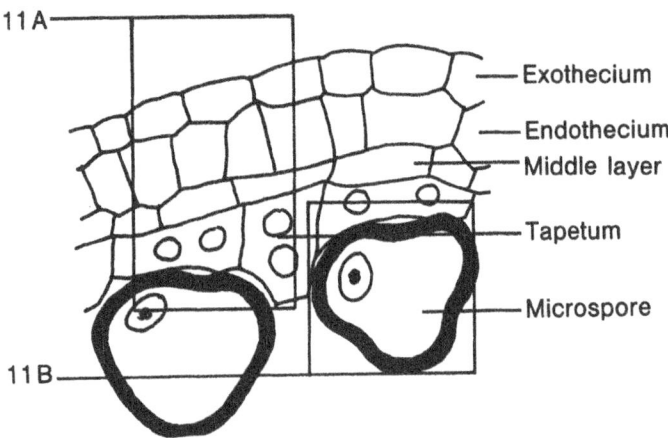

Plate 11A. Transverse section through an anther of *Brassica oleracea* L., showing the various tissues at the early free microspore stage (compare with plate 1A); the exothecium, the endothecium, the remains of the middle layers, the tapetum cells, and the microspores. Most cells of the middle layers have already disappeared. The strongly enlarged tapetum cells are still in their original position, although the cell walls bordering the microspores are dissolving. x 4,000.

Plate 11B. Portion of an anther of *Helianthus annuus* L. at the early free microspore stage. The microspores already have a well developed exine and a very thick lamellate structured endexine. The tapetal cells have lost their cell walls, became amoeboid, and invaded the locule. x 6,000.

(Plate 11A courtesy M Rougier, Lyon; Plate 11B courtesy F Ciampolini, Siena).

MS
T
N
Nu
T
ML
En
Ex
V
A

T
MS
N
B

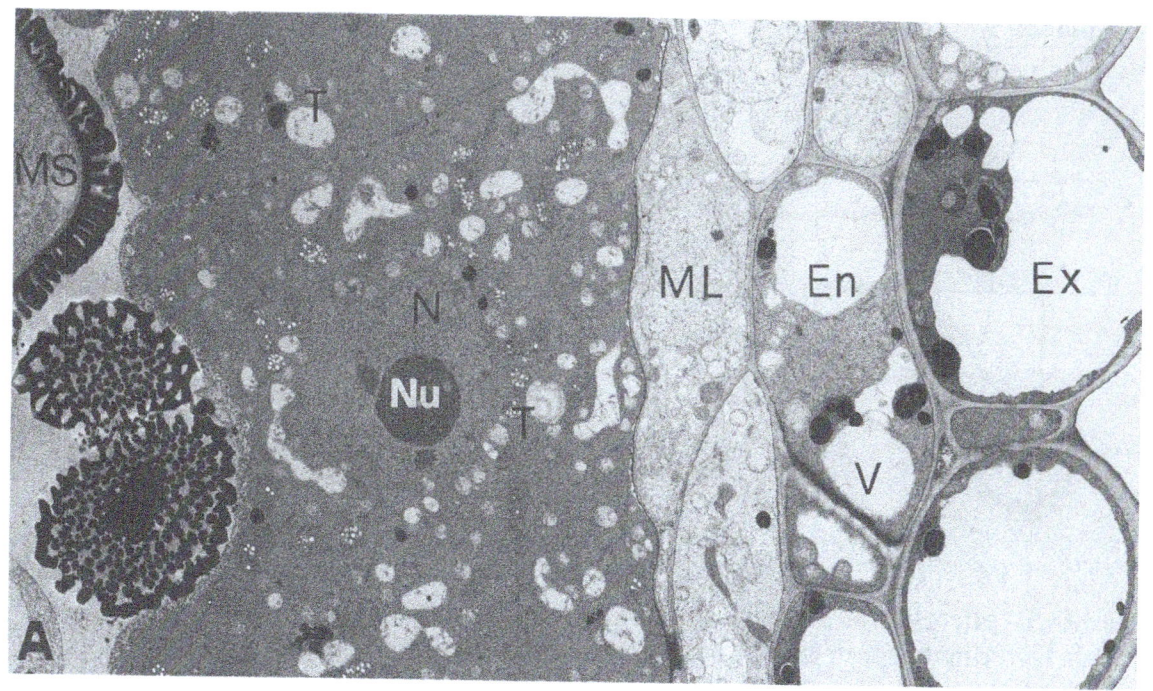

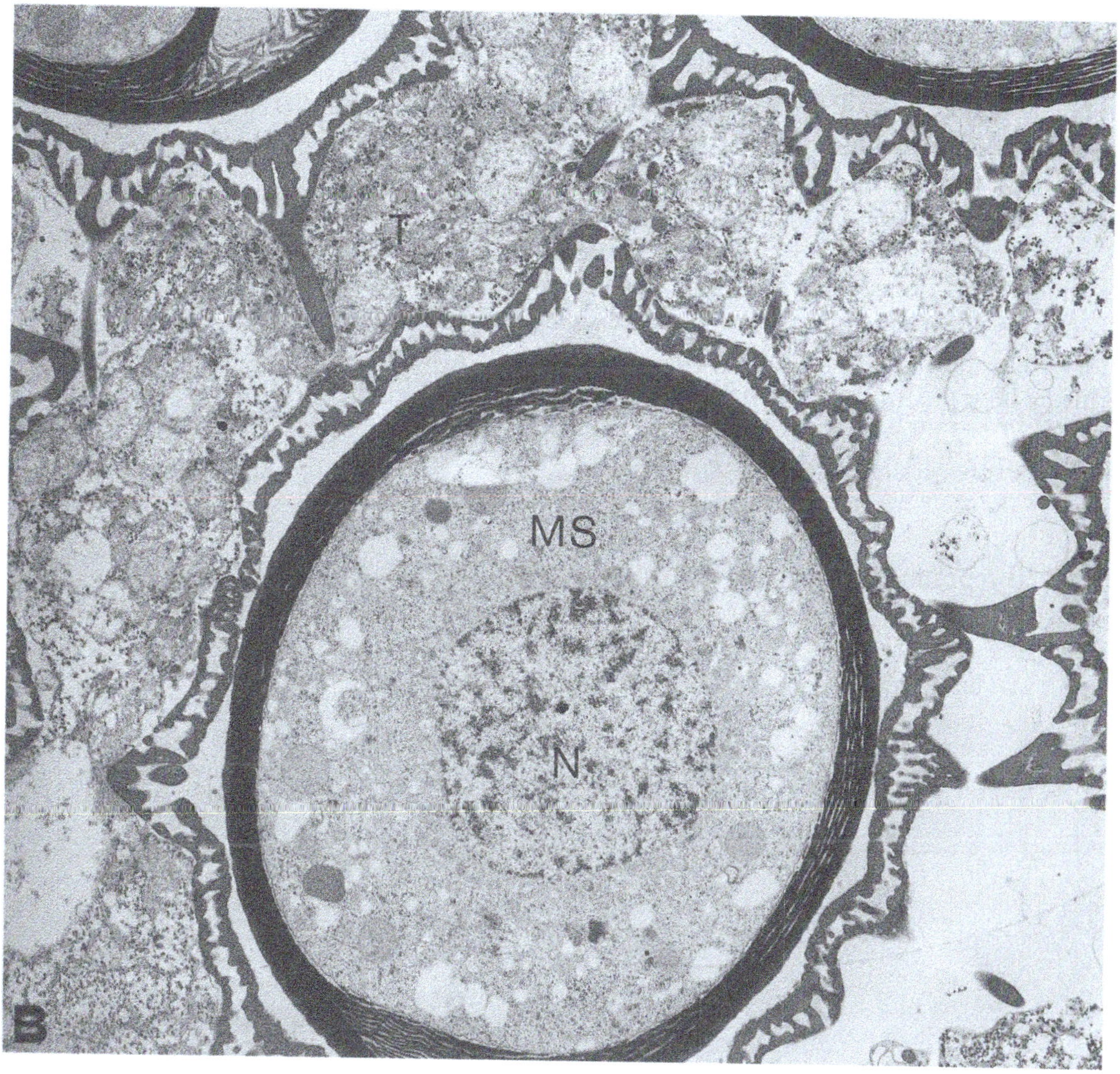

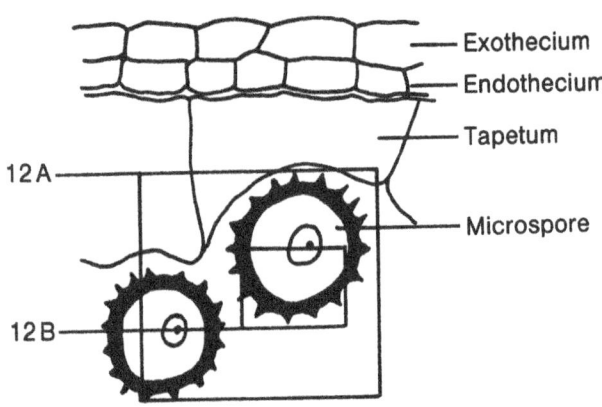

Plate 12A/B. Microspores and tapetum of *Catananche caerulea* L. at the early free microspore stage. The material has been chemically fixed, freeze-fractured and critical-point dried for scanning electron microscopy.

A. The tapetal cells still form a complete layer surrounding the microspores, but they are about to become amoeboid. Numerous stacked cisternae of endoplasmic reticulum are apparent in the tapetum cells (arrows). x 4,000.

B. Enlarged portion of a microspore and neighbouring tapetum cell. At this stage the microspore wall is already composed of an exine and an intine. Note the thickness and lamellate structure of the intine near the future germination aperture (asterisk). Note the presence of strands of sporopollenin precursors (arrows) produced by the tapetum cells. x 20,000.

(Plate 12B reproduced by permission from Barnes S H, Backmore S (1988) Ann Bot 62: 615-623).

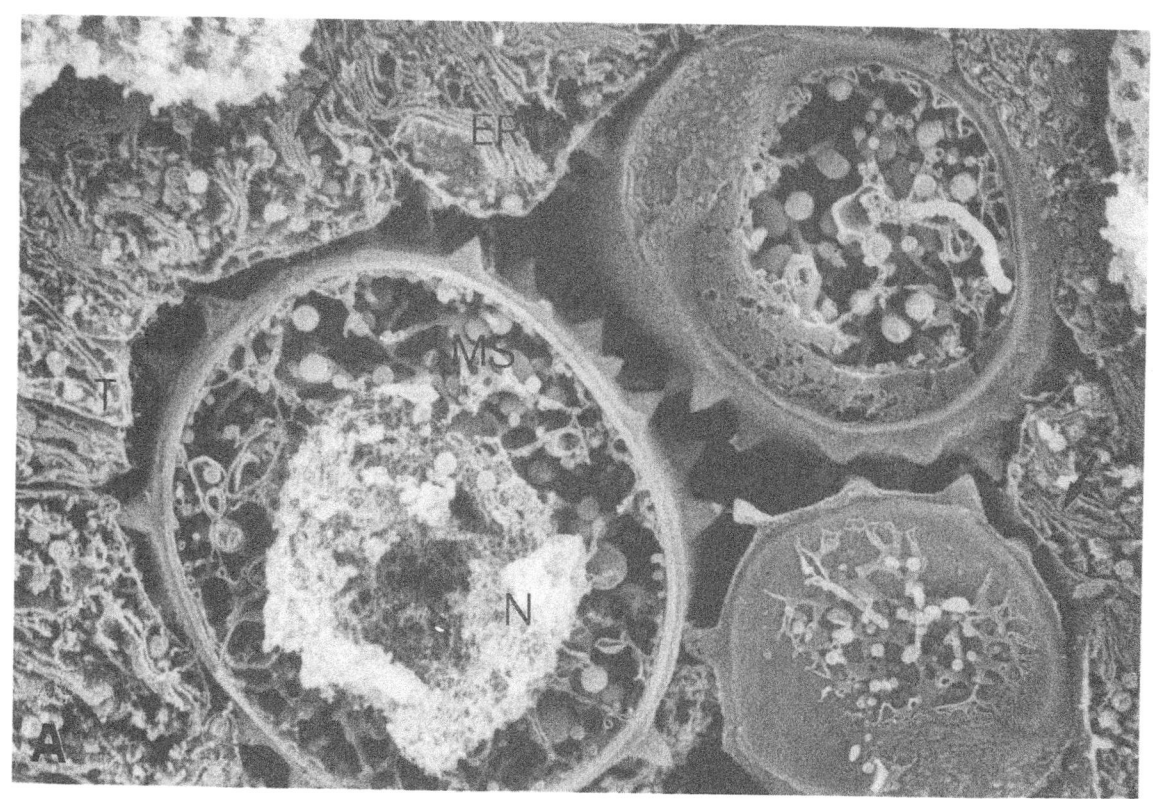

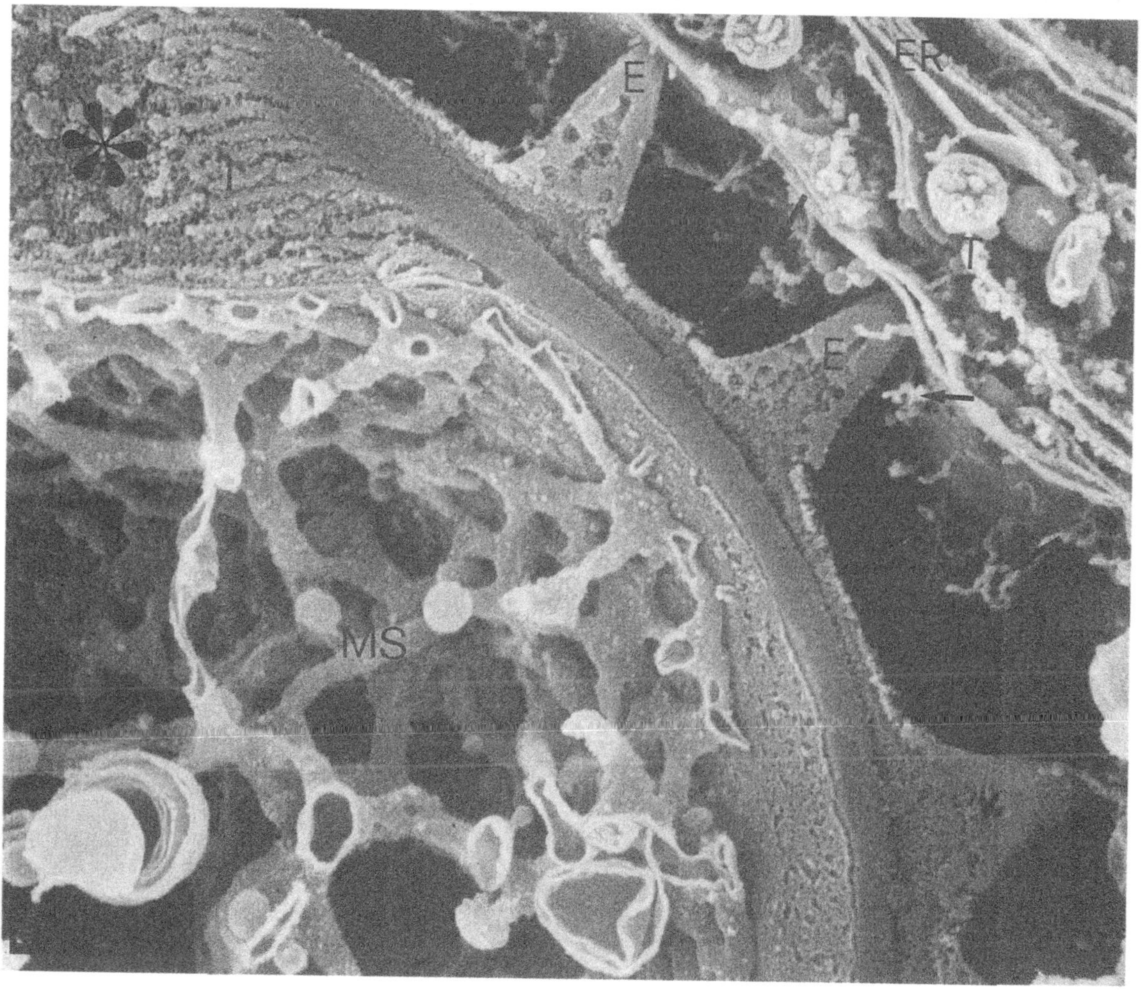

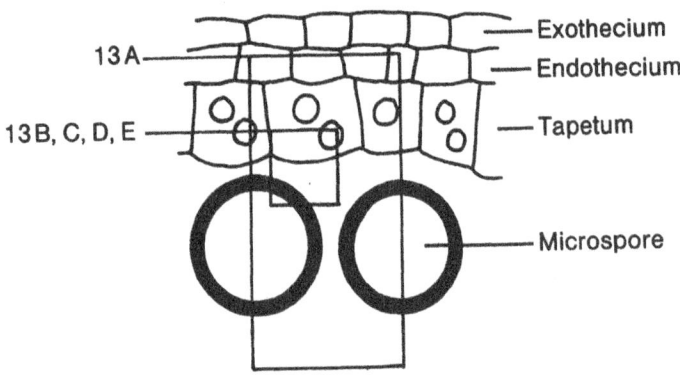

Plate 13A. Part of an anther of *Olea europaea* L. at free microspore stage. The tapetum is of the secretory type, which remains cellular, although the inner tangential and radial walls disappear. x 3,000.

Plate 13B. Enlarged portion of a tapetal cell of *Olea europaea* L. The tapetal cell contains stacked endoplasmic reticulum. x 12,000.

Plate 13C. Portion of a secretory tapetum cell of *Arbutus unedo* L. Materials are released from the tapetum cells by exocytosis (arrow). At the surface of the cell a pro-orbicule is forming (double arrow). x 25,000.

Plate 13D. Portion of an almost ripe anther of *Parietaria judaica* L. The secretory tapetum has already degenerated. The loculus is delimited by orbicles (Ubisch bodies) (arrows), produced by the tapetum. x 7,000.

Plate 13E. Portion of an almost ripe anther of *Forsythia viridissima* Lindl. Each tapetal cell produces a ball of lipophylic material (asterisk), surrounded by orbicules. Later, the material is deposited on the pollen grains as pollenkitt. x 8,000.

(Plates 13A,E courtesy E Pacini, Siena).

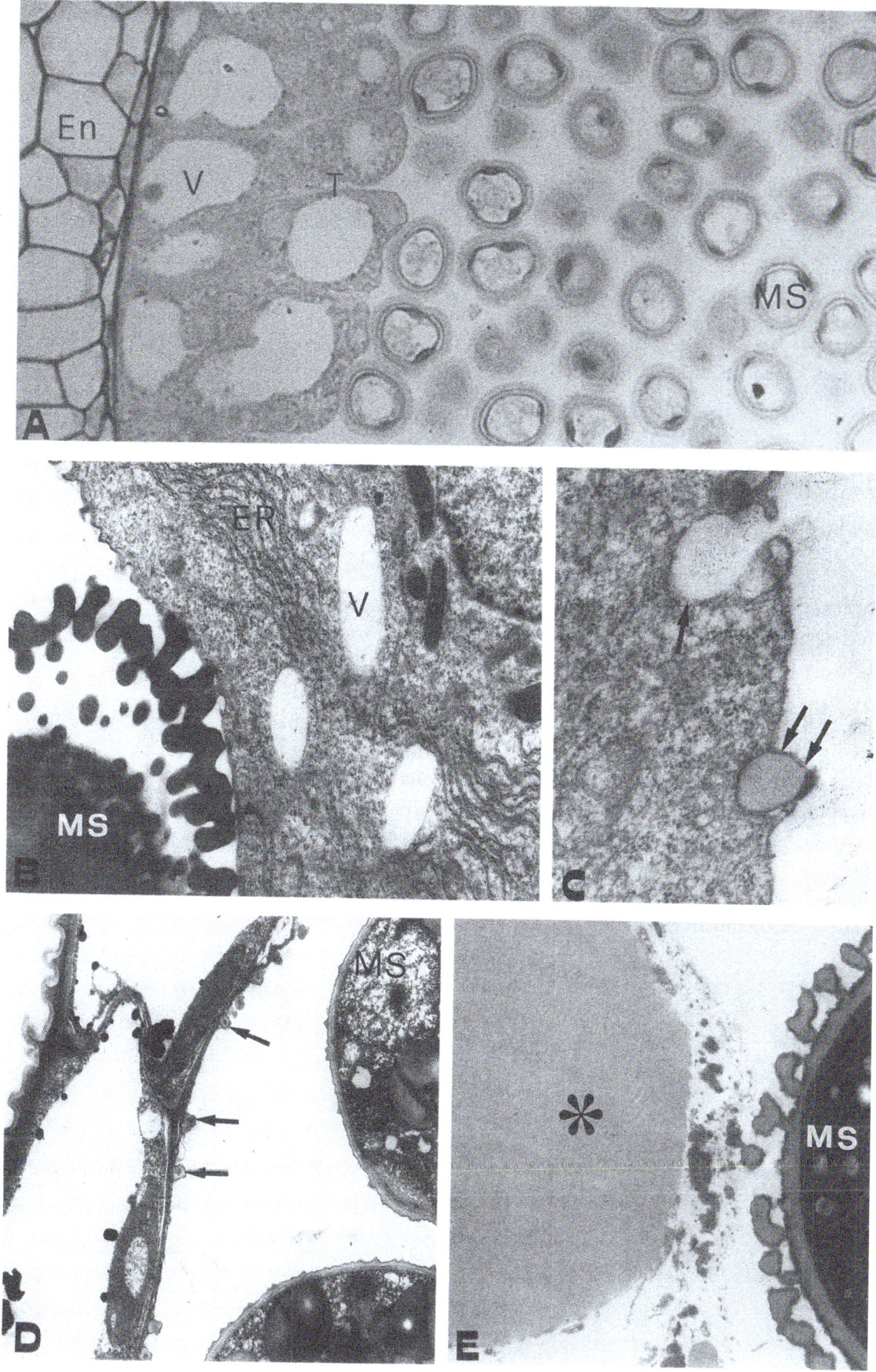

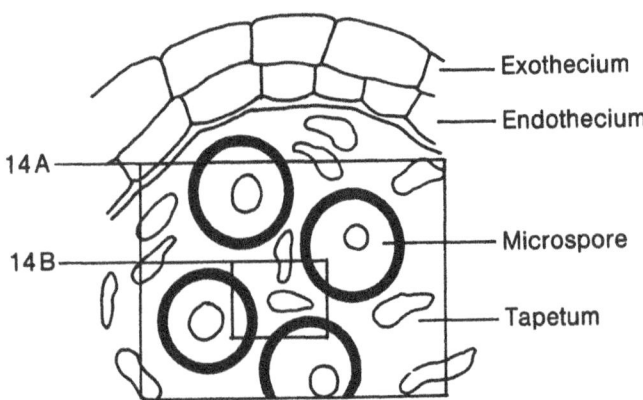

Exothecium

Endothecium

14A

Microspore

14B

Tapetum

Plate 14A. Part of an anther of *Arum italicum* Mill. at early free microspore stage. The tapetal cells have fused to form a syncytium. This type of tapetum is called periplasmodial. The tapetal cytoplasmic mass invades the locule, and surrounds and adheres to the developing microspores. Two zones are present in the tapetal cytoplasm: one adjacent to the microspores, and a second one in which vacuoles and nuclei are present. x 1,500.

Plate 14B. Enlarged portion of an anther of *Arum italicum* Mill. showing the ultrastructure of the periplasmodial tapetum. The tapetal cytoplasm near the microspores contains mainly ribosomes, vesiculate endoplasmic reticulum and microtubules (arrows). The microtubules are aligned parallel to the exine surface. x 20,000.

(Plates 14A,B courtesy E Pacini, Siena).

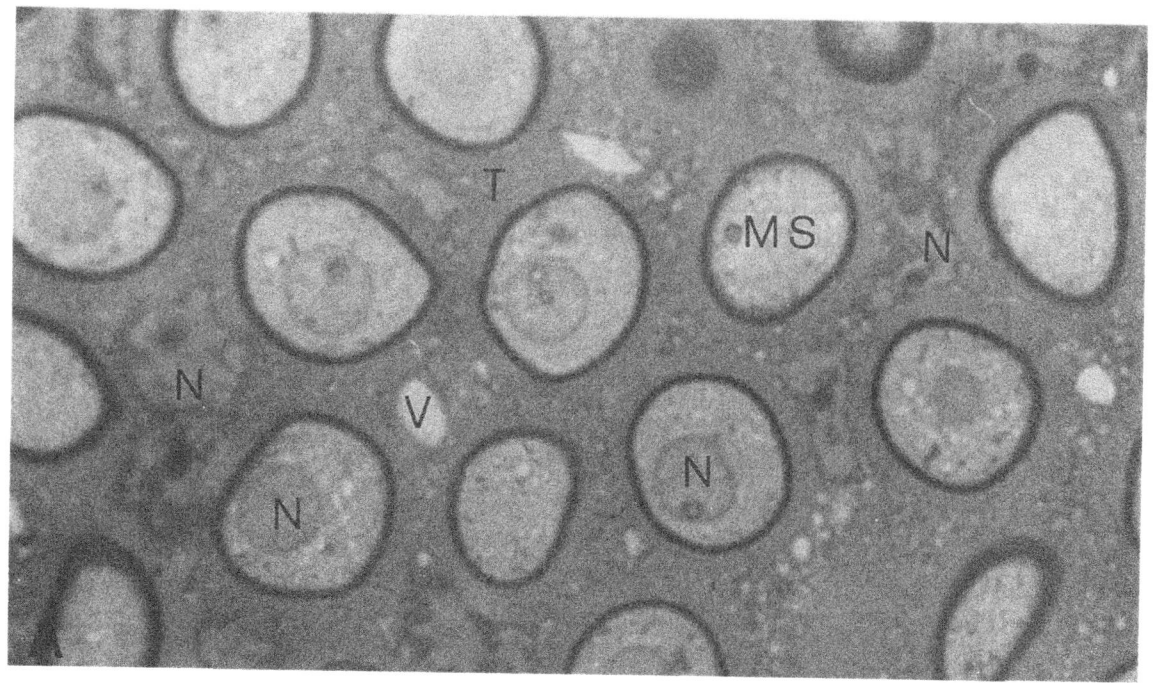

B

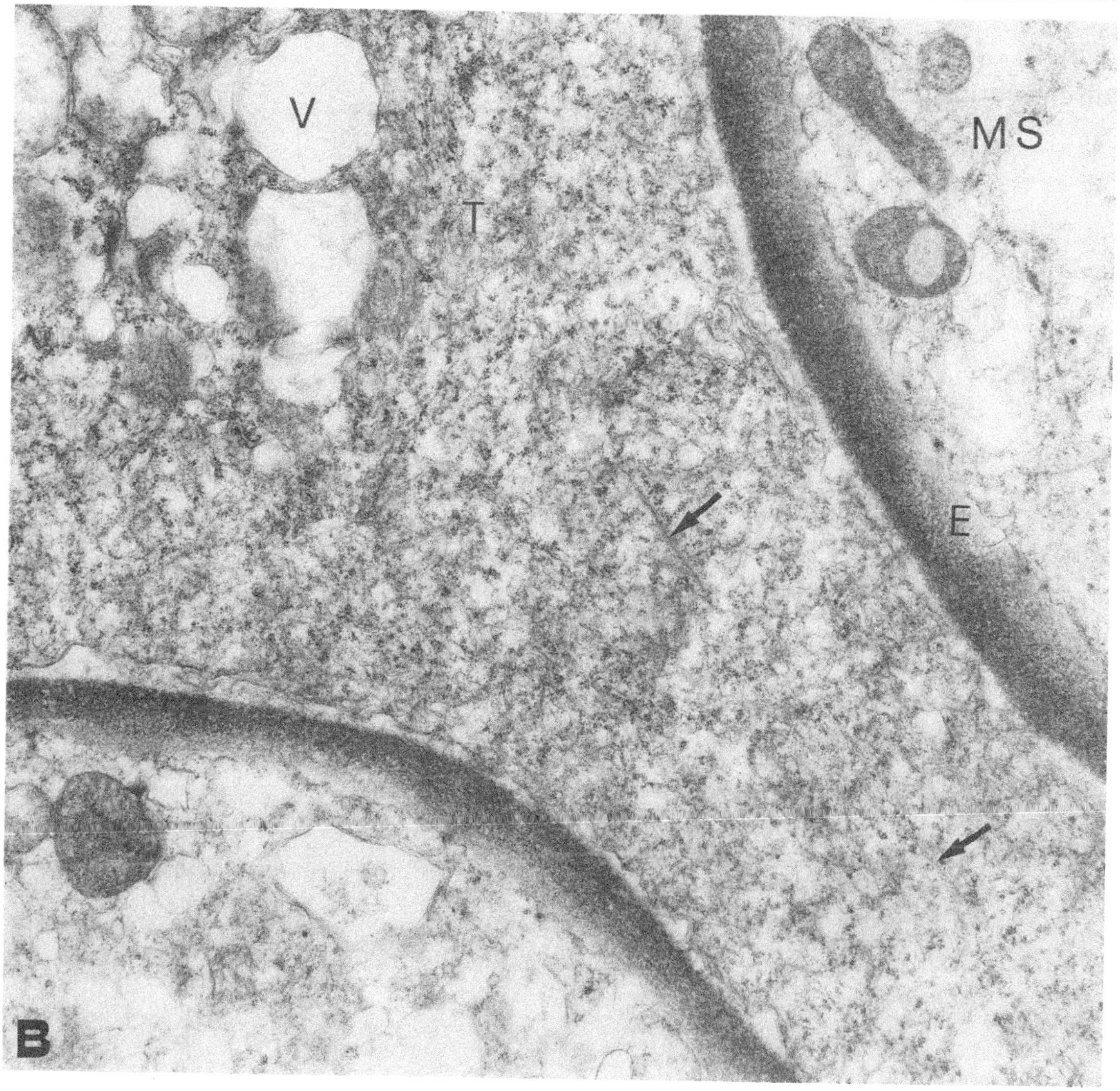

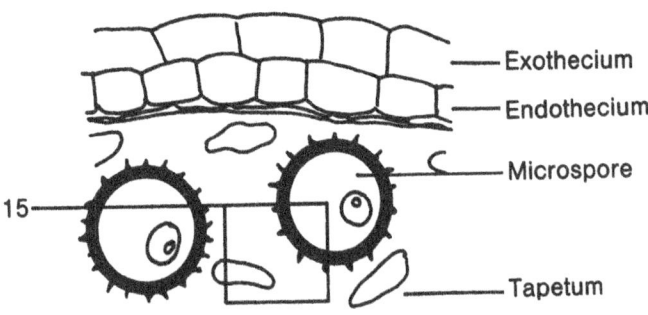

Plate 15. Enlarged portion of a tapetum cell of *Helianthus tuberosus* L., close to a young microspore. The tapetal cell has an amoeboid character and is closely appressed to the exine of the microspores. One of the exine baculae can be seen deeply inside the tapetum cell, without disrupting the plasma membrane (arrows). x 27,000.

(Plate 15 courtesy F Ciampolini, Siena).

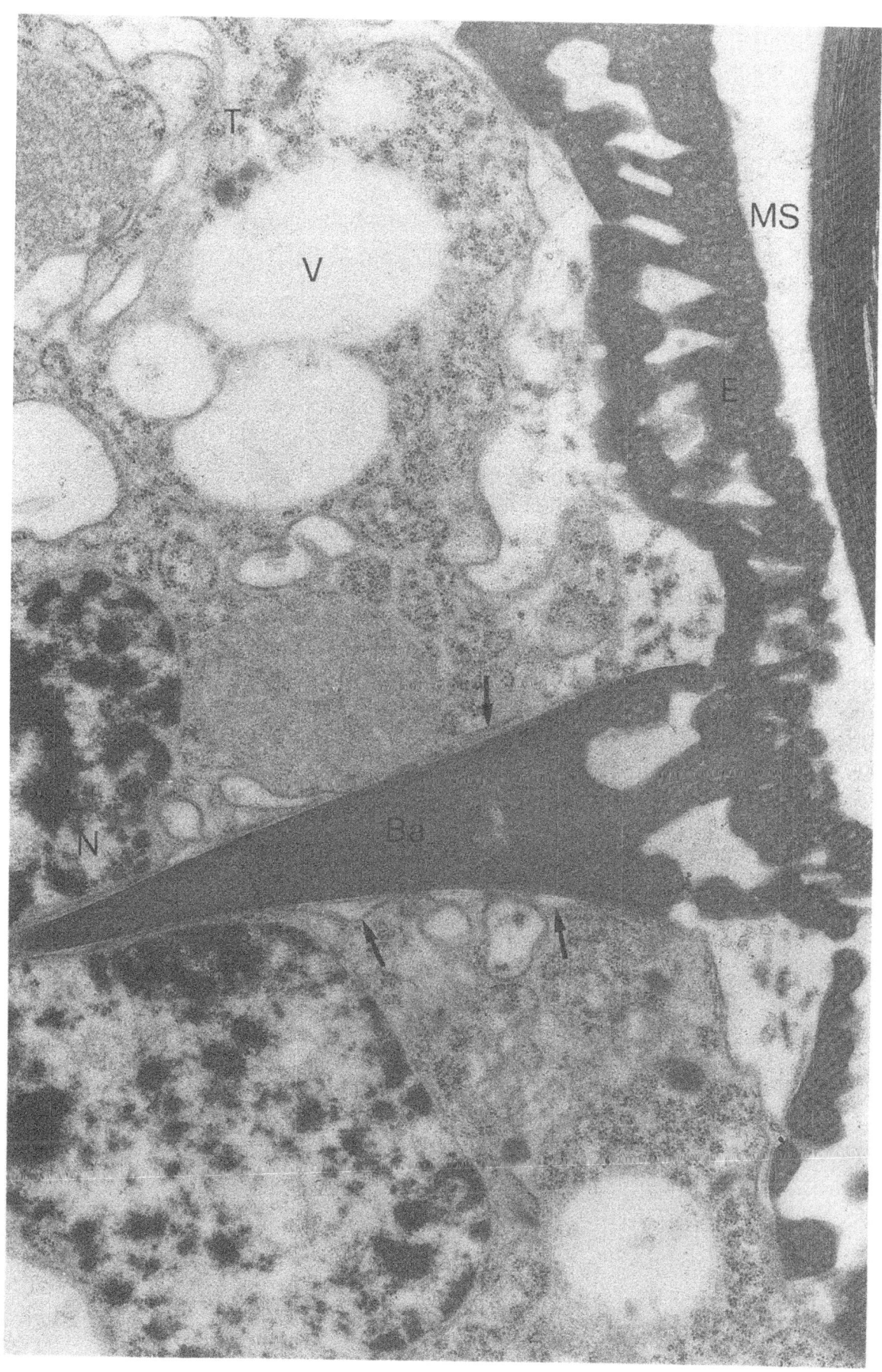

47

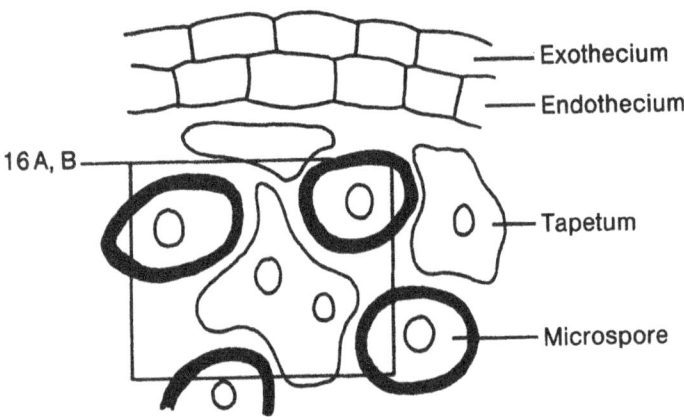

Plate 16 A, B. Transverse sections through normal (**A**) and cytoplasmic male sterile anthers (**B**) of *Impatiens walleriana* Hook f. In cytoplasmic male sterile anthers of *Impatiens walleriana*, development is normal until the microspore stage. In fertile anthers the microspores enlarge after their release from the tetrads. The formation of the exine proceeds, and the cells become vacuolate. The tapetal cells are binucleate and amoeboid. In the cytoplasmic male sterile anthers, the tapetal cells enlarge strongly and become multinucleate. However, the microspore enlargment does not occur and vacuoles are not formed. Ultimately the microspores become plasmolyzed and they degenerate. x 3,000.

(Plates 16A,B reproduced by permission from Van Went J (1981) Acta Soc Bot Pol 50: 249-252).

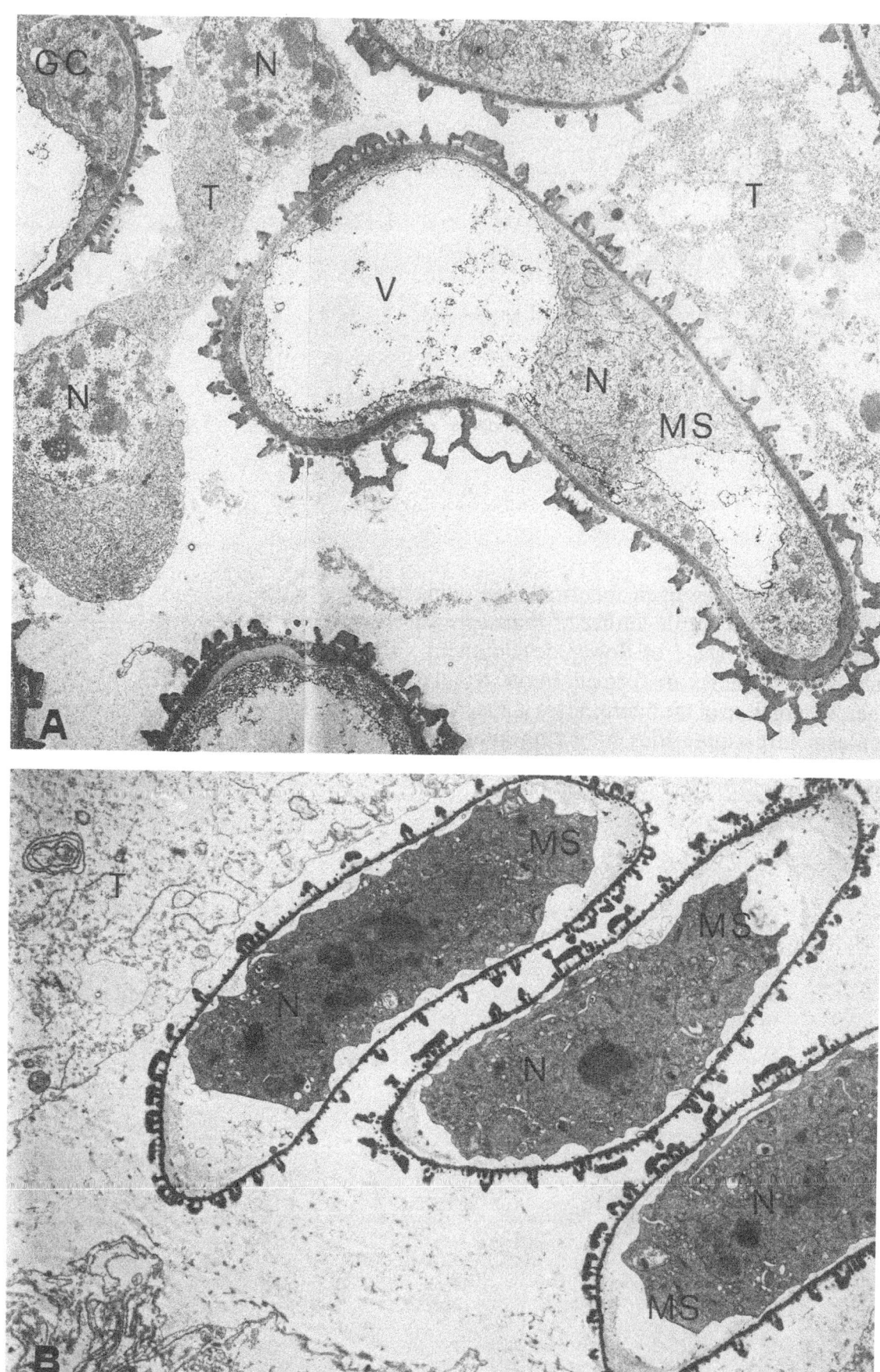

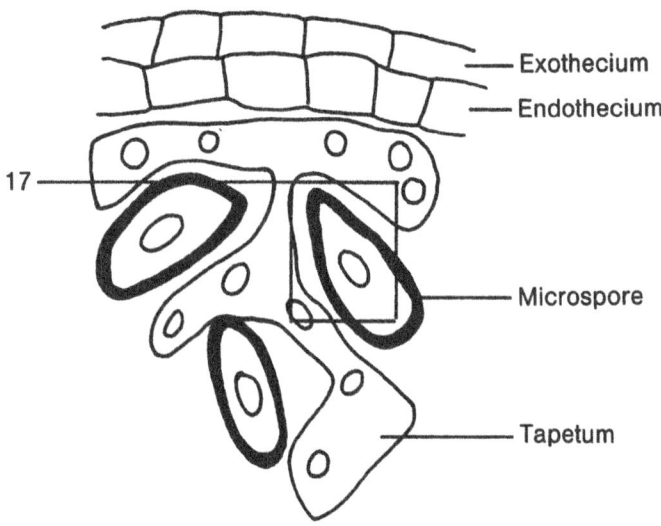

Plate 17. Detailed view of an abnormal developing microspore in a cytoplasminc male sterile anther of *Impatiens walleriana* Hook f. Normally, at this stage of flower development, the microspore enlarges, and vacuoles are formed. In the cytoplasmic male sterile anther, vacuolation of the microspores is apparently blocked. In the cytoplasm large quantities of membranes are formed, which develop into large aggregates (asterisk). x 12,000.

(Original J Van Went).

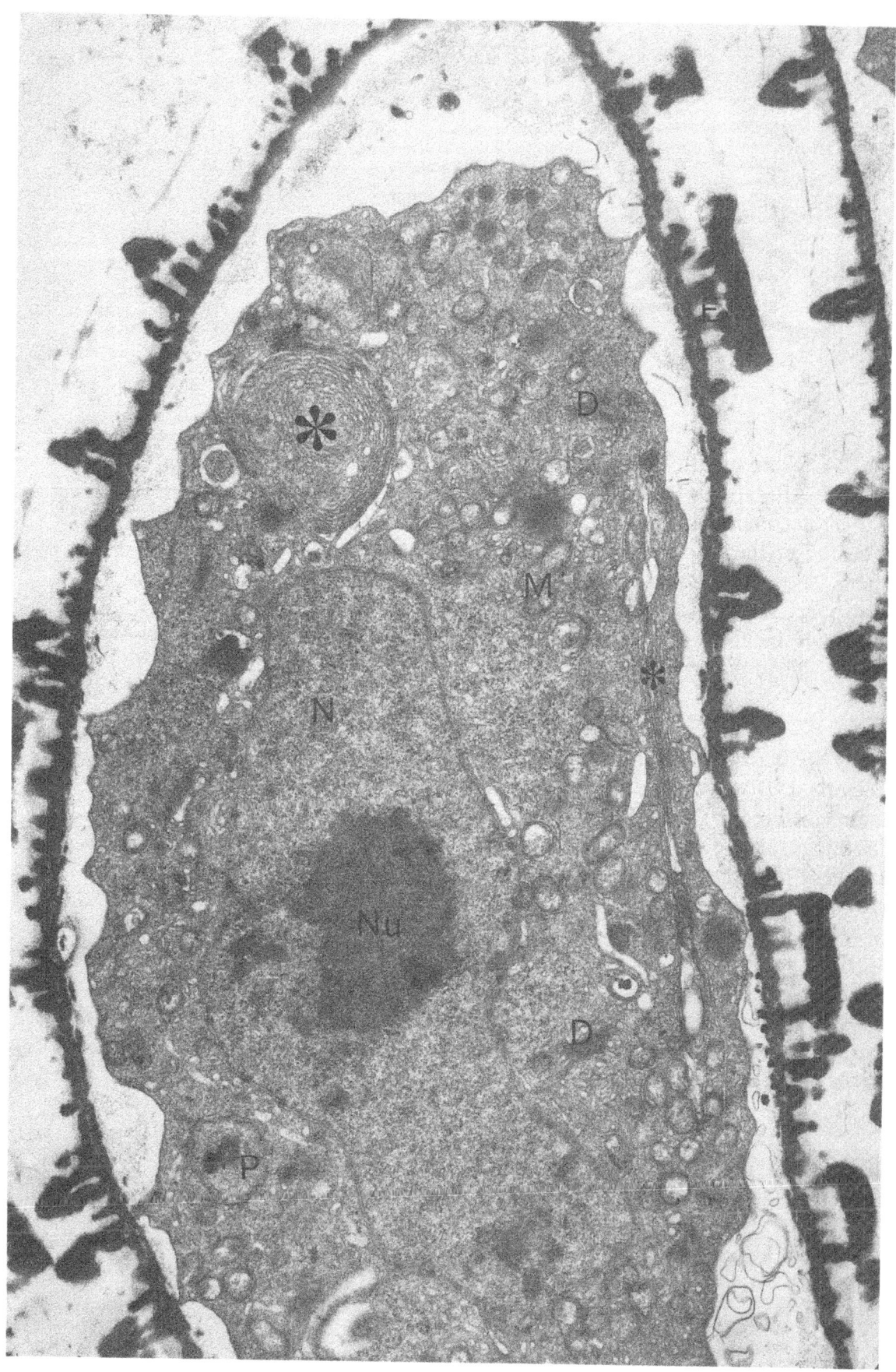

51

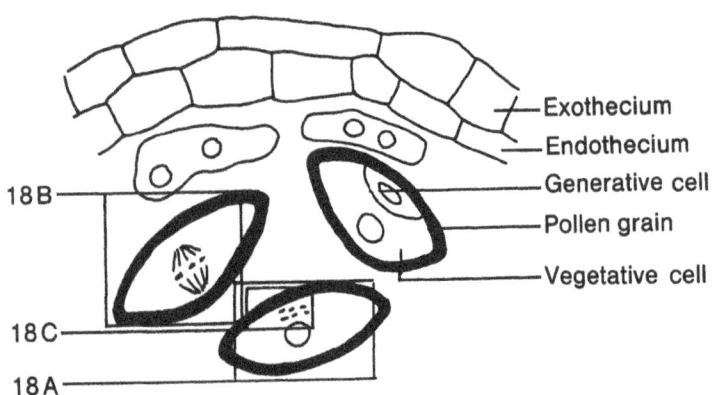

Exothecium
Endothecium
Generative cell
Pollen grain
Vegetative cell

18B
18C
18A

Plate 18A. Microspore of *Impatiens walleriana* Hook f. at the onset of mitosis. After its release from the tetrad the microspore enlarges, which is accompanied by the formation of vacuoles, and the thickening of the exine. At the onset of mitosis the microspore nucleus takes an acentral position in the mid region of the microspore. At the same time the plastids and several other cytoplasmic inclusion also gather in the mid-region, but at the opposite side of the microspore (arrows). x 3,500.

Plate 18B. Microspore of *Impatiens walleriana* Hook f. at metaphase of mitosis. During mitosis the plastids and many other organelles remain clustered in that region of the microspore which will later form the vegetative cell of the pollen grain. Like the nucleus, the metaphase plate has also an acentral position in the microspore (arrows). x 4,000.

Plate 18C. Enlarged view of the clustered organelles in a microspore of *Impatiens walleriana* Hook f. at the onset of mitosis. Plastids and mitochondria are closely packed, as are intermingling cisterns of endoplasmic reticulum (arrows). x 20,000.

(Plates 18A, C reproduced by permission from Van Went J (1984) Theor Appl Genet 68: 305-309).

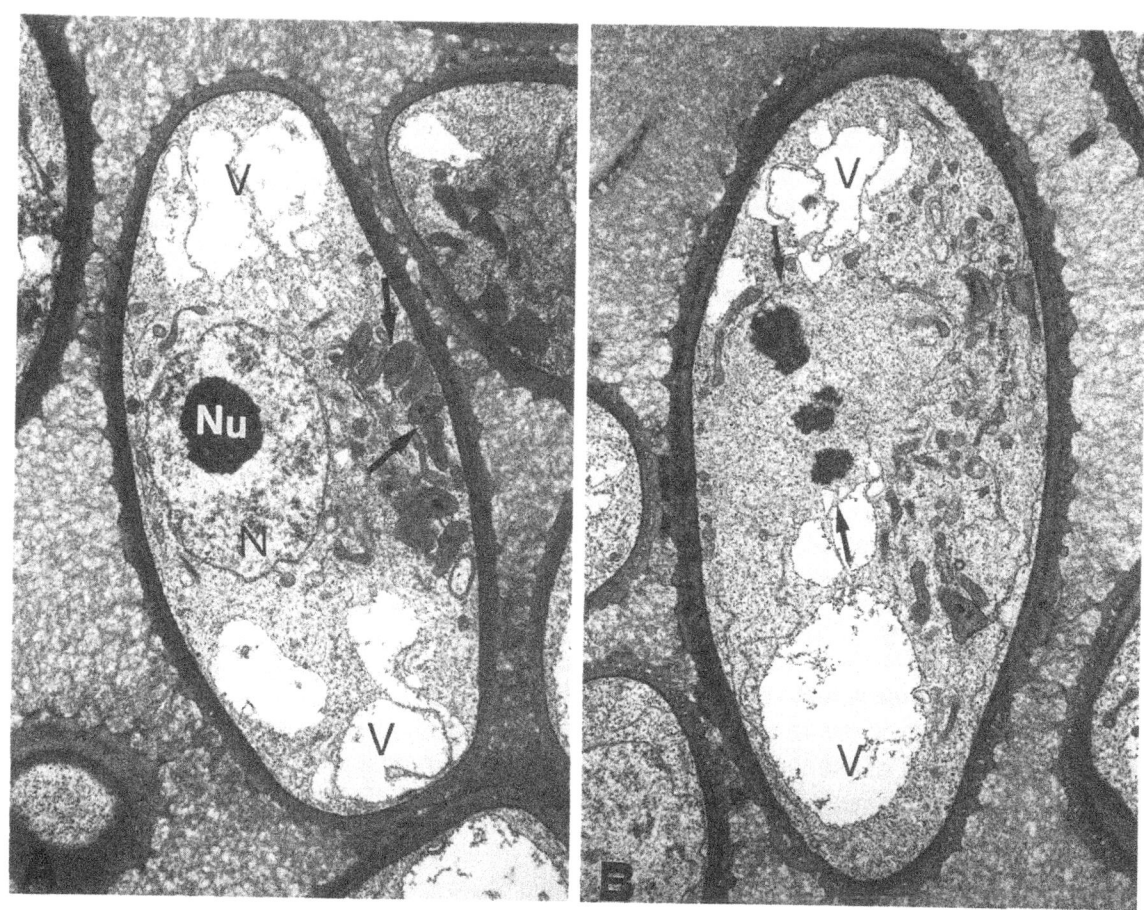

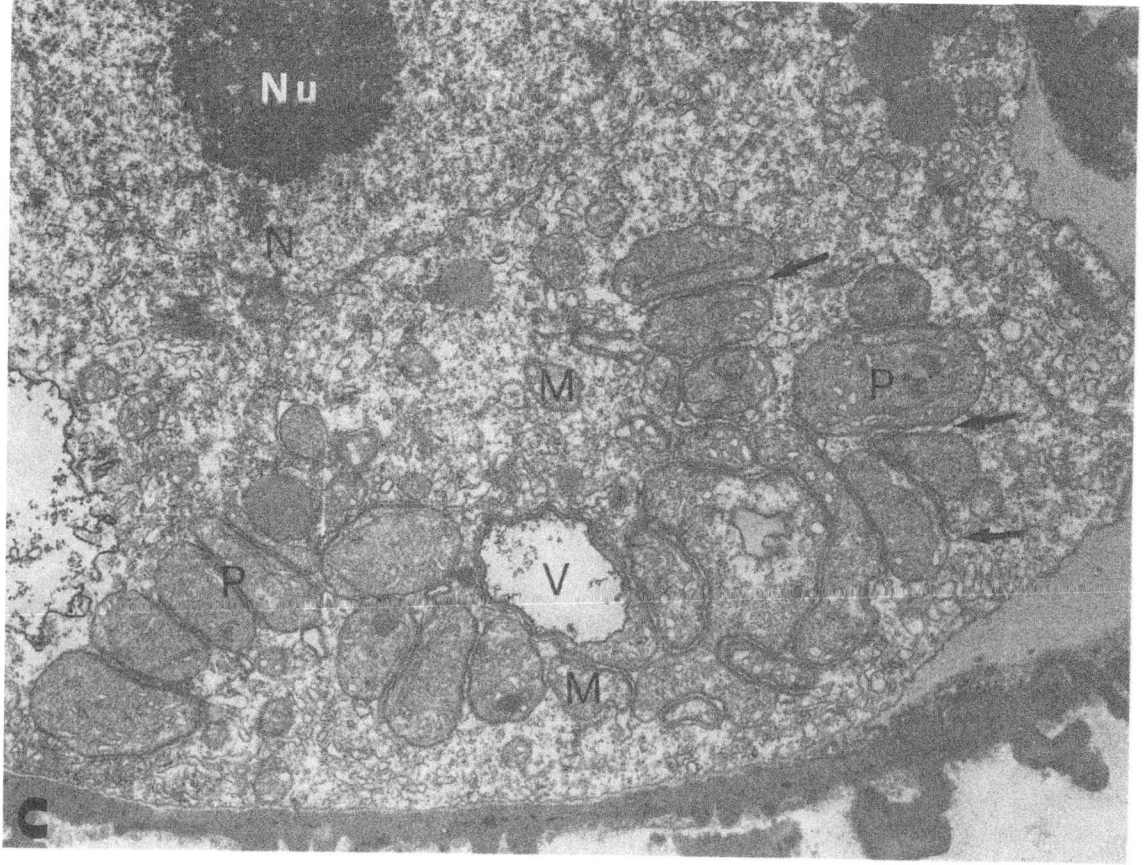

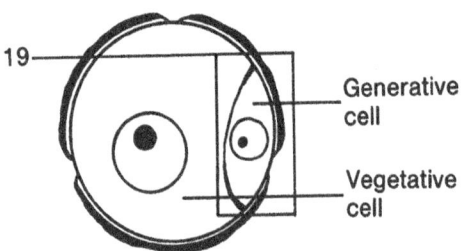

Plate 19. Generative cell of the pollen grain of *Euphorbia dulcis* L., just after its formation. The newly formed generative cell is lenticularly shaped and attached to the intine. It is much smaller than the vegetative cell, because of the unequal division of the microspore. The generative cell has only few organelles and no plastids, as a result of the clustering of most organelles and all plastids during the microspore division in such a position that they are directly transferred to the vegetative cell. At this stage of development the wall of the generative cell is pecto-cellulosic and continuous with the intine. x 20,000.

(Original M Cresti).

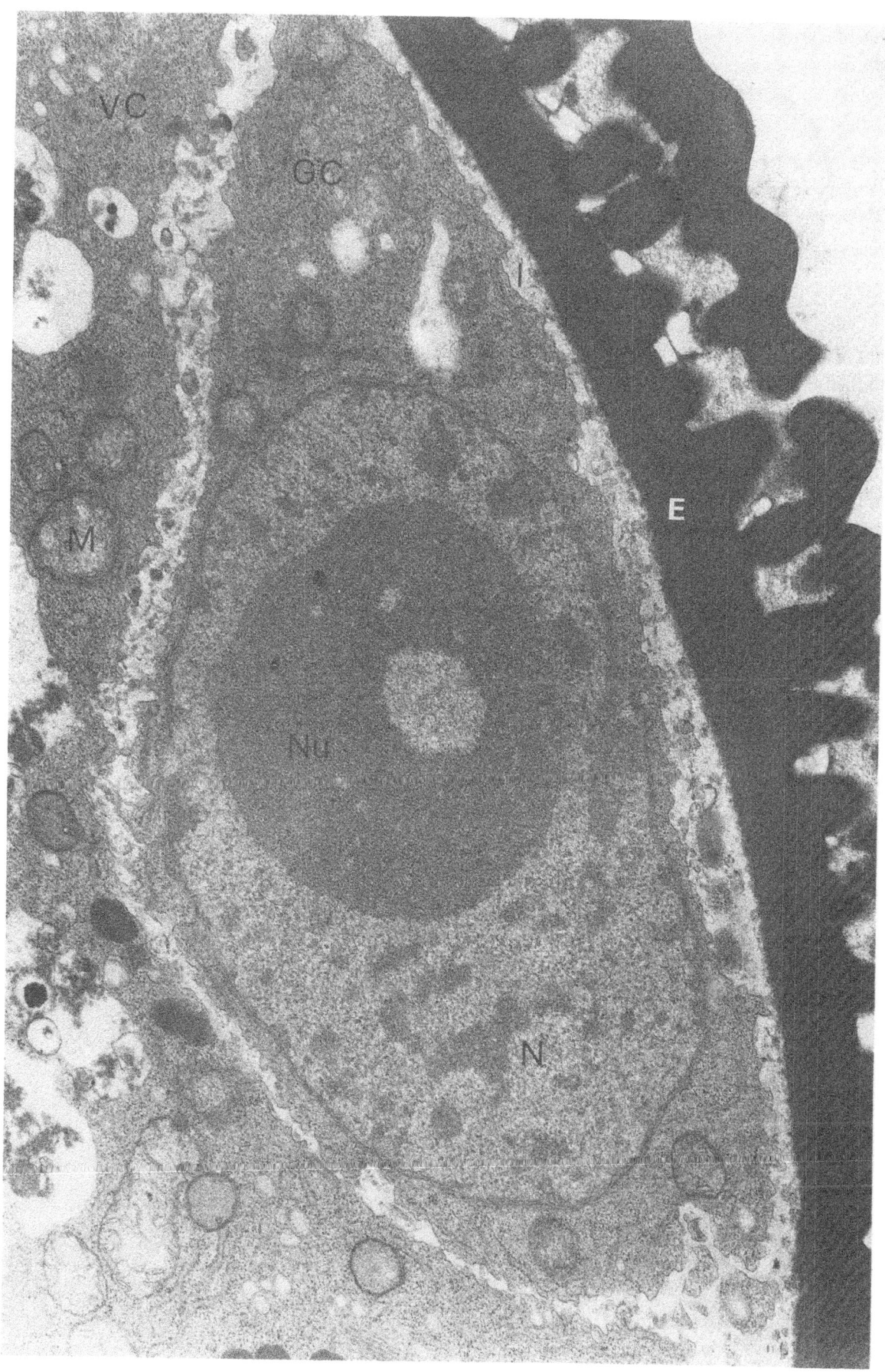

55

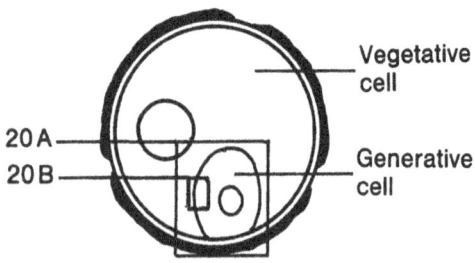

Plate 20A. Generative cell of *Euphorbia dulcis* L., during its detachment from the intine. After its formation the generative cell gradually detaches from the intine by a centripetally constriction at the attachment site (arrows). Simultaneously the shape of the generative cell changes to spherical. Note the accumulation of spherosomes/lipid bodies in the vegetative cell at the surface of the generative cell. x 33,500.

Plate 20B. Enlarged portion of plate 20A, showing the interface between the generative cell and the vegetative cell of *Euphorbia dulcis* L.. During the detachment of the generative cell from the intine, the morphology of the wall in between the two cells changes. The two plasma membranes become very straight and positioned very closely to each other. The generative cell develops a cytoskeleton of which the microtubules near the plasma membrane are very evident (arrows). x 60,000.

(Plates 20 A,B courtesy M Murgia, Siena).

E
M
P
GN
Nu
GC
VC
L
L
VN
Nu
A
VC
L
L
GC
ER
M
M
B

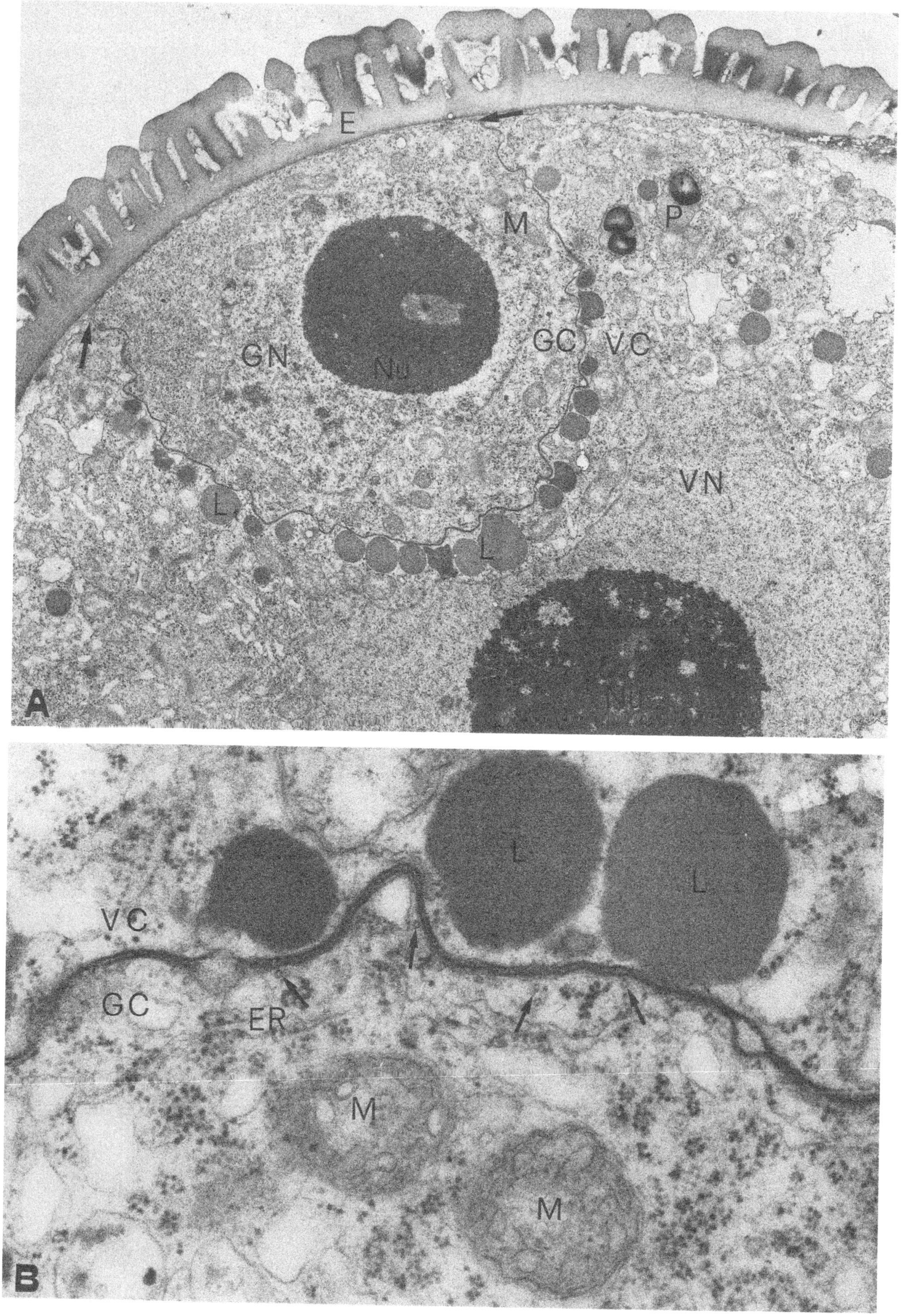

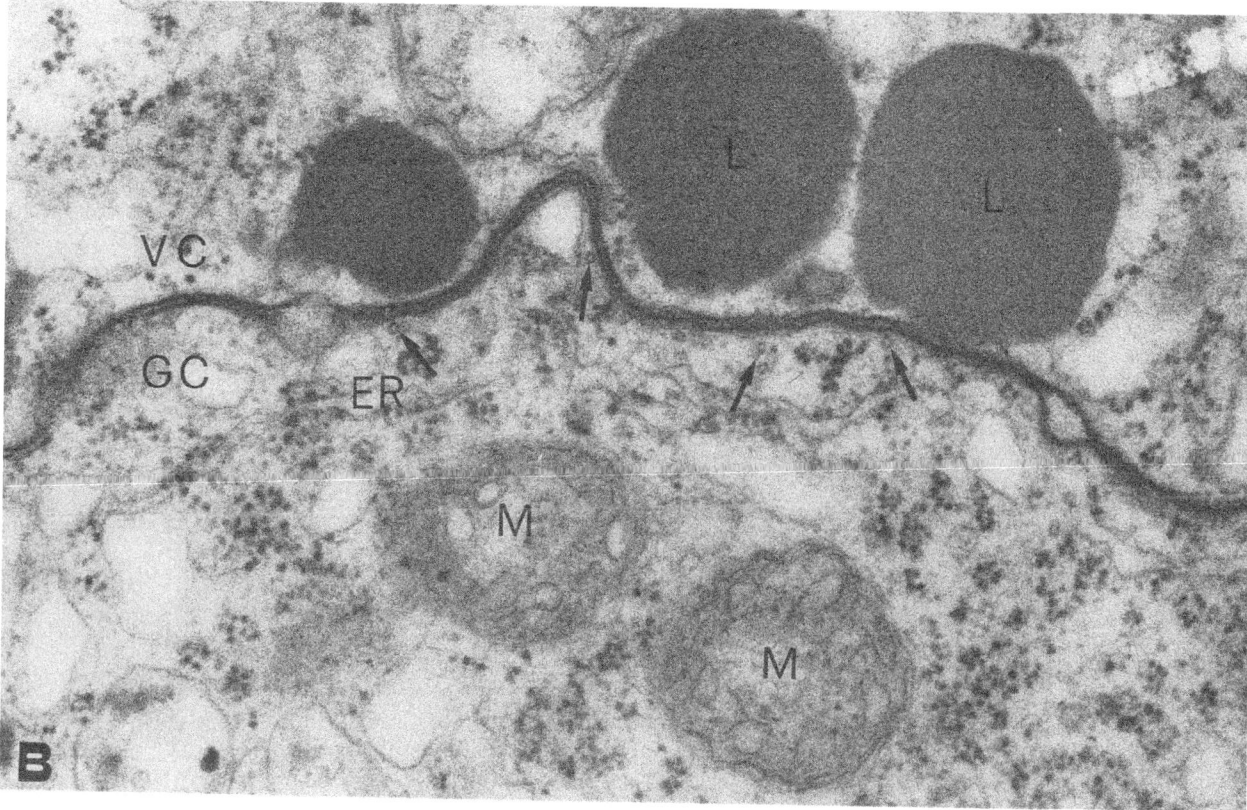

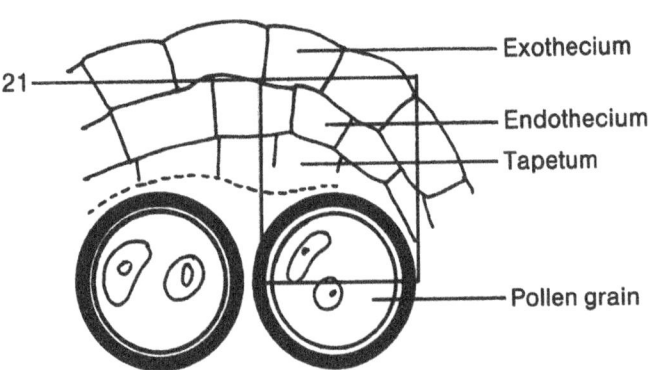

Plate 21. Anther organization of *Euphorbia dulcis* L. at the bicellular pollen stage, with free, spherical generative cell. Directly after its detachment from the intine the generative cell is spherical and moves to a position more deeply inside the vegetative cell. At this stage of development the tapetum starts to degenerate. At the surface of the degenerating tapetum deposits of sporopollenin are present (arrows). x 18,600.

(Plate 21 courtesy M Murgia, Siena).

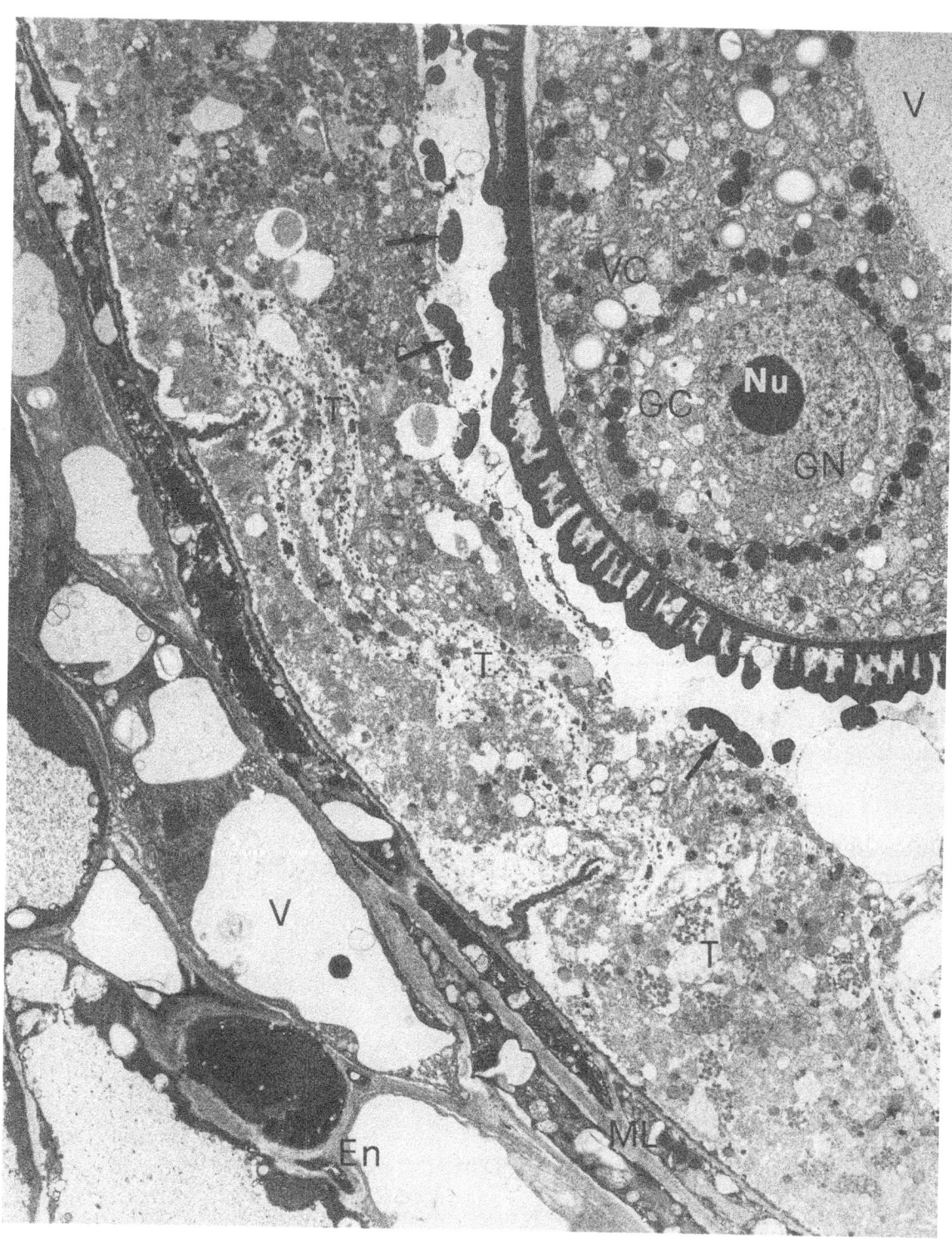

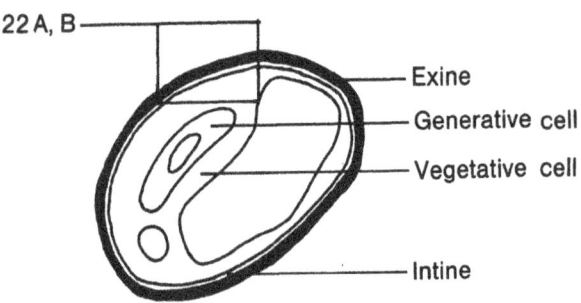

22 A, B
Exine
Generative cell
Vegetative cell
Intine

Plate 22A. Portion of a maturating pollen grain of *Secale cereale* L.. During maturation the pollen grain becomes completely filled with cytoplasm, and its large vacuole disappears. During this stage the intine shows local thickenings, resulting in an enlarged plasma membrane surface. This structure probably acts as a "transfer" cell wall, involved in the uptake of nutrients needed for plasma synthesis. x 20,000.

Plate 22B. Detailed view of the wall projections of the intine of the maturating *Secale cereale* L. pollen grain. x 41,500.

(Plate 22A courtesy M Charzynska, Warsaw; Plate 22B reproduced by permission from M Charzynska et al (1990) Protoplasma 158: 26-32).

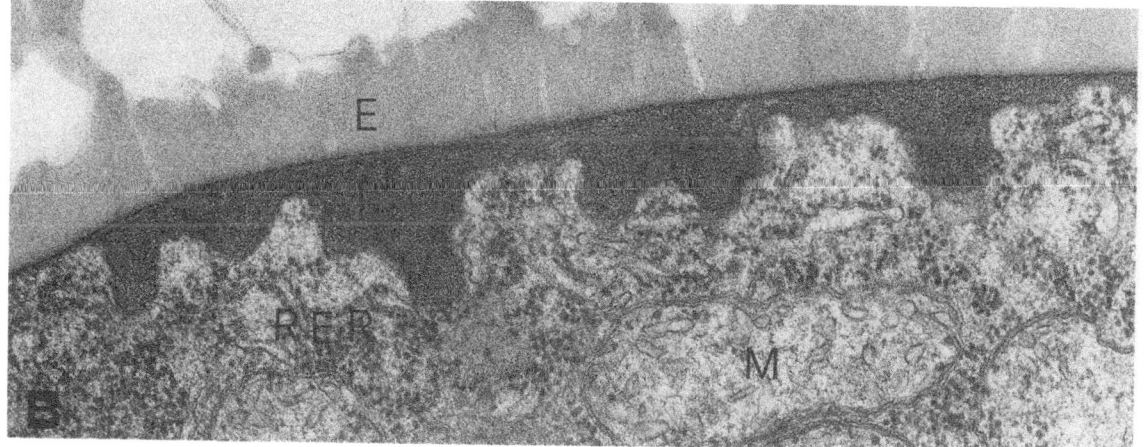

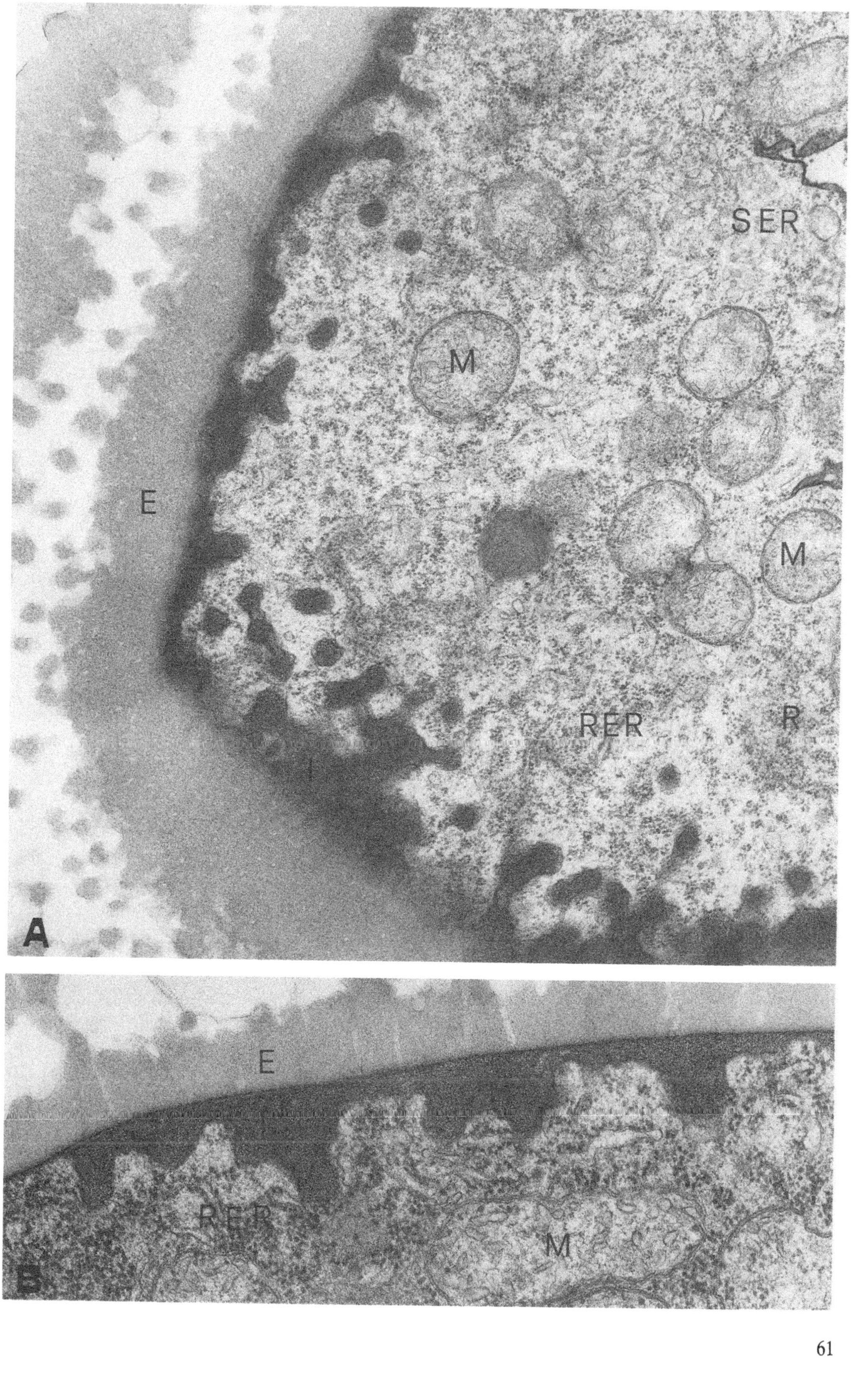

61

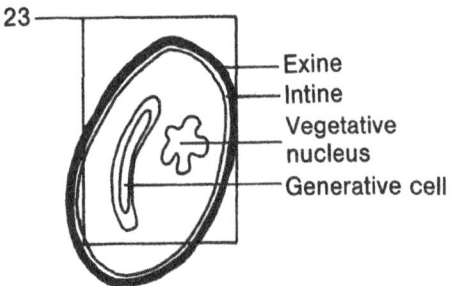

Exine
Intine
Vegetative
nucleus
Generative cell

Plate 23. Mature bicellular pollen of *Tradescantia virginiana* L.. In *Tradescantia*, as in many other angiosperm species, the mature pollen grain is composed of two cells, a large vegetative cell and a much smaller generative cell. At the mature stage, the vegetative cell is enclosed by a bilayered wall, the outer exine and the inner intine. The exine consists of sporopollenin, whereas the intine is a normal pecto-cellulosic wall. The vegetative cell is completely filled with cytoplasm, and only smaller vacuoles are present. The nucleus of the vegetative cell is very irregular in shape with many protrusions and invaginations. The generative cell has an elongated shape and its cytoplasm shows a much simpler ultrastructure then the vegetative cell. The nucleus of the generative cell has a highly condensed appearance, indicating that is already preparing for the mitotic division leading to the formation of the two sperm cells. x 6,200.

(Plate 23 courtesy F Ciampolini, Siena).

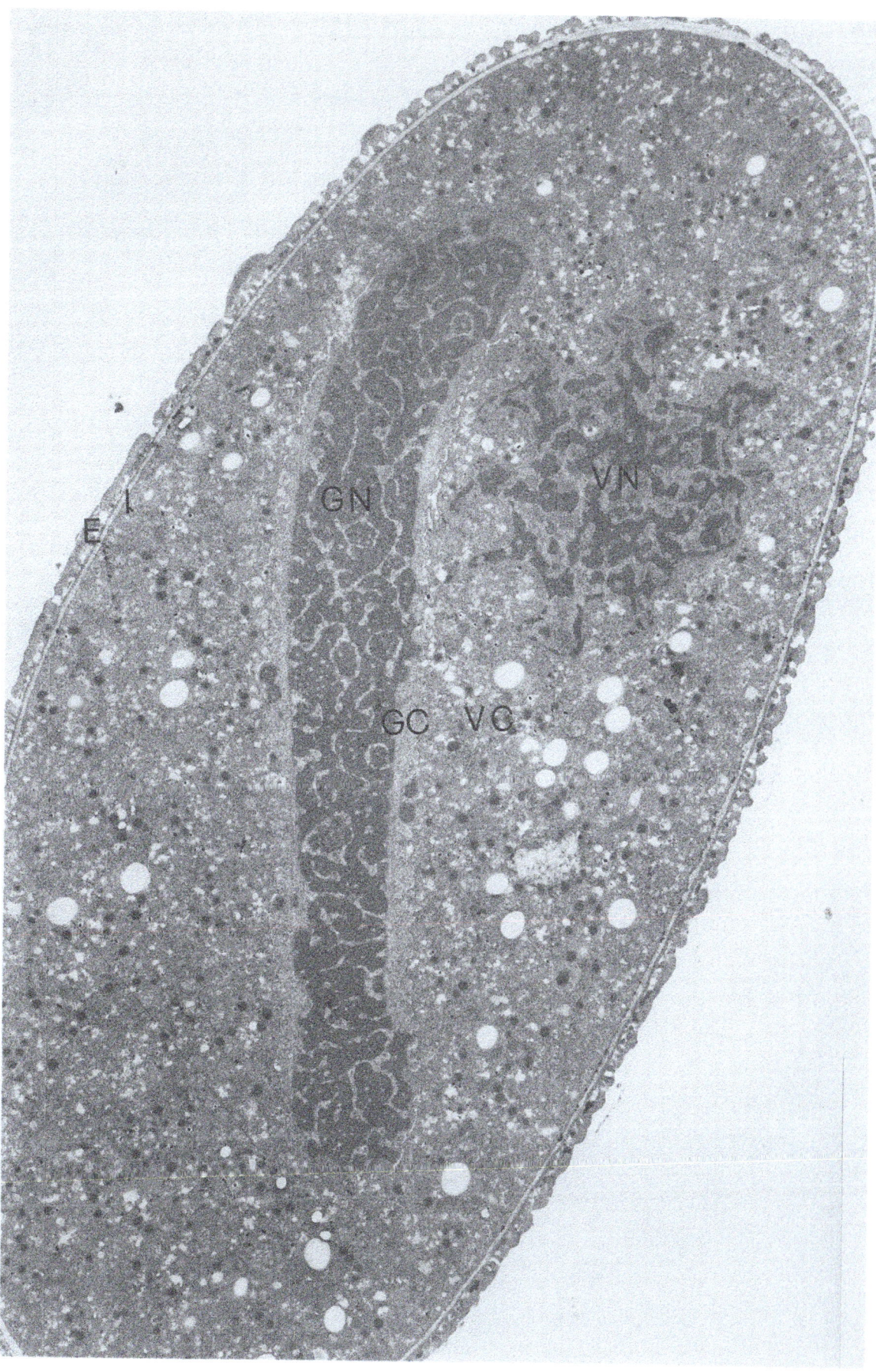

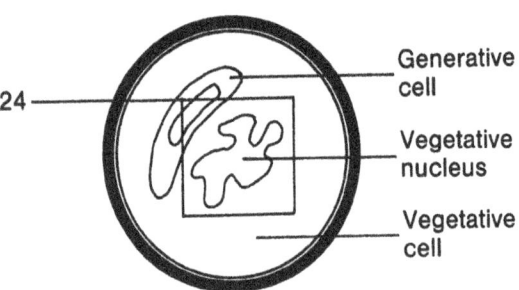

24

Generative cell

Vegetative nucleus

Vegetative cell

Plate 24. Ultrastructure of the vegetative cell of *Amaryllis bella-donna* L.. At this stage the cytoplasm shows a complex organization and composition. The nucleus of the vegetative cell is very irregular in outline. Two trans-section of the generative cell are present in this section of the pollen grain. x 8,500.

(Plate 24 courtesy F Ciampolini, Siena).

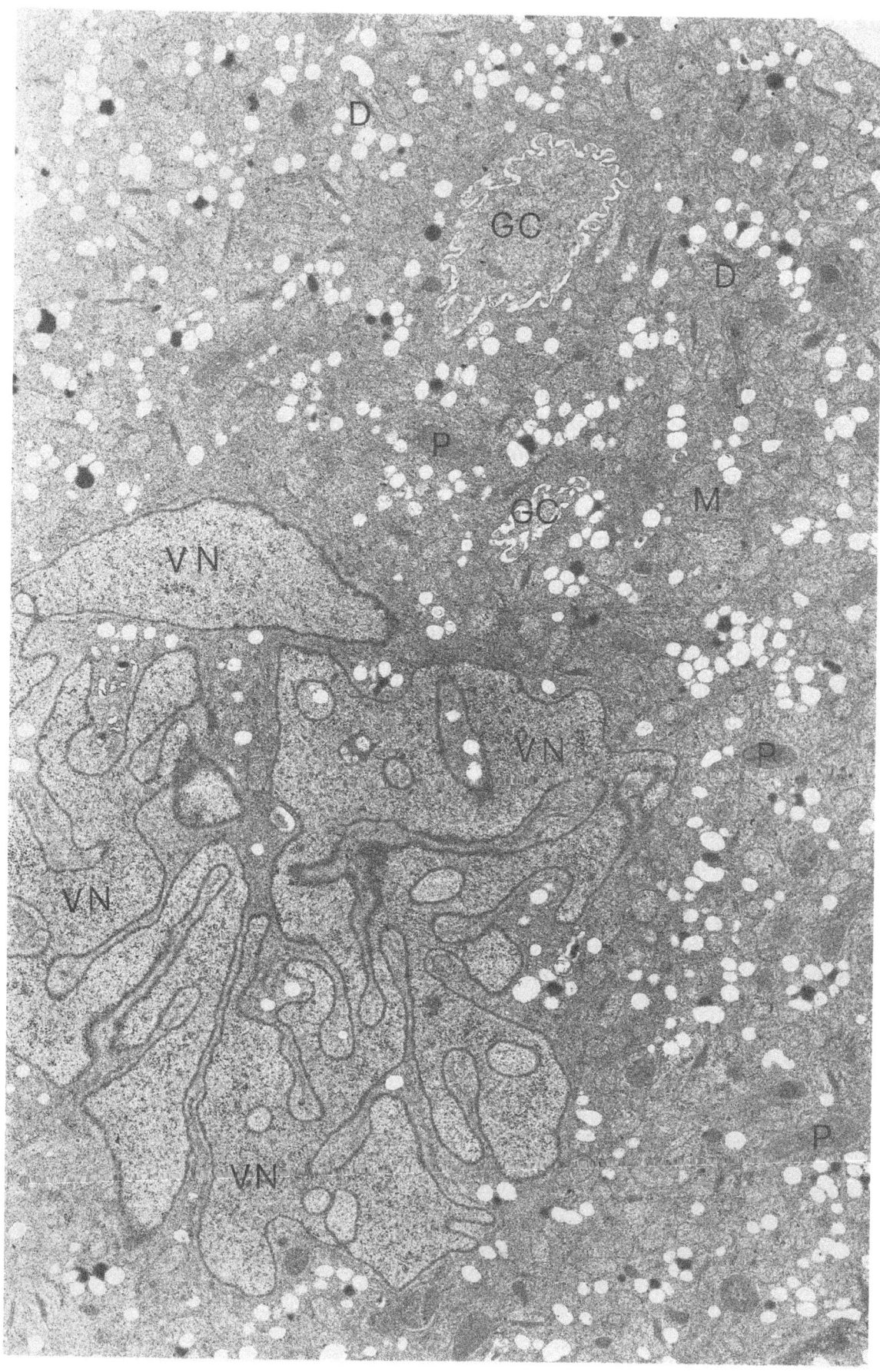

65

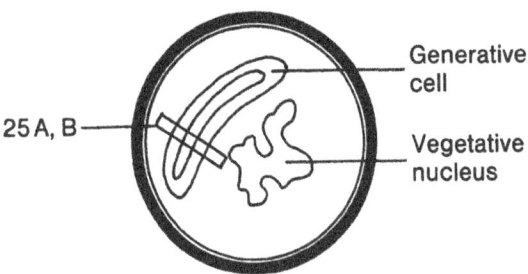

Plate 25A.. Transverse section through the generative cell of *Amaryllis bella-donna* L. at the mature pollen stage. The generative cell is spindle shaped with the elongated nucleus positioned in the central region of the cell. In cross section the generative cell has a round profile with a strongly undulating outline. In the peripheral cytoplasmic region large numbers of microtubules (arrows) can be observed, running parallel to the outer surface and the long axis of the cell. The two neighbouring plasma membranes of the vegetative and the generative cell run fairly parallel to each other, but there is considerable and varying space in between them. The generative nucleus is highly heterochromatic with many electron-dense particles in the euchromatic regions. The cytoplasmic constitution is relatively simple. x 23,000

Plate 25B. Enlarged portion of a generative cell of *Nicotiana alata* Link & Otto, showing the bundles of microtubules which run parallel to the long axis of the generative cell. The plasma membranes of the generative and the vegetative cell are separated from each other by considerable space (arrowheads). This space is thought to be an artifact caused by the chemical fixation, since it is not observed in cryo-fixed, freeze-substituted material. x 40,000.

(Plates 25 A,B courtesy F Ciampolini, Siena).

66

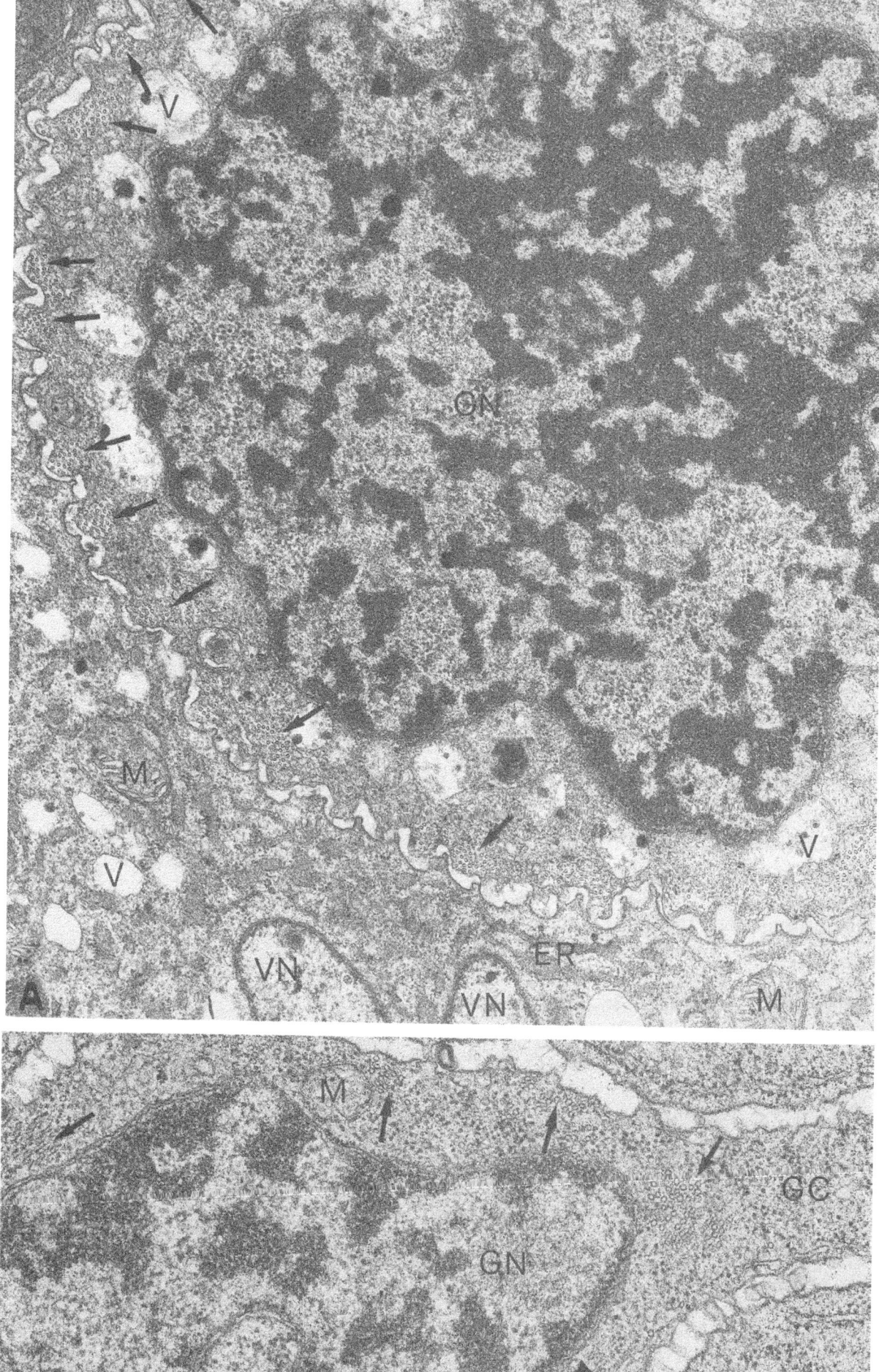

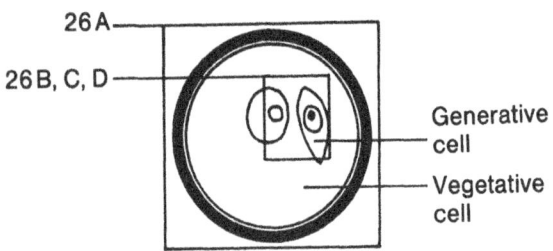

Plate 26. Mature bicellular pollen of *Papaver dubium* L.. The material has been cryo-fixed and freeze-fractured for observation with low-temperature high-resolution field emission scanning electron microscopy.

A. Two fractured pollen grains showing the position and morphology of the spindle shaped generative cells. The outer surface shows ridges running parallel to the long axis. x 1,400.

B. Enlarged portion showing the partially fractured vegetative nucleus, and a transverse fractured generative cell. The membrane of the irregularly shaped vegetative nucleus has numerous, randomly distributed nuclear pores (arrow). In cross-section the undulating outline of the generative cell is clearly visible (arrowheads). In the cytoplasm of the generative cell only few organelles are present, whereas the vegetative cytoplasm contains many inclusions. The generative cell is positioned close to the vegetative nucleus. x 10,000.

C. Enlarged portion, showing the vegetative nucleus and the adjacent generative cell with its undulating surface. x 10,000.

D. Enlarged portion of plate 26A, showing the spindle shape of the generative cell and the ridges at its surface. x 5,000.

(Plates 26A-D reproduced by permission from A Van Aelst et al (1989) Acta Bot Neerl 38: 25-30).

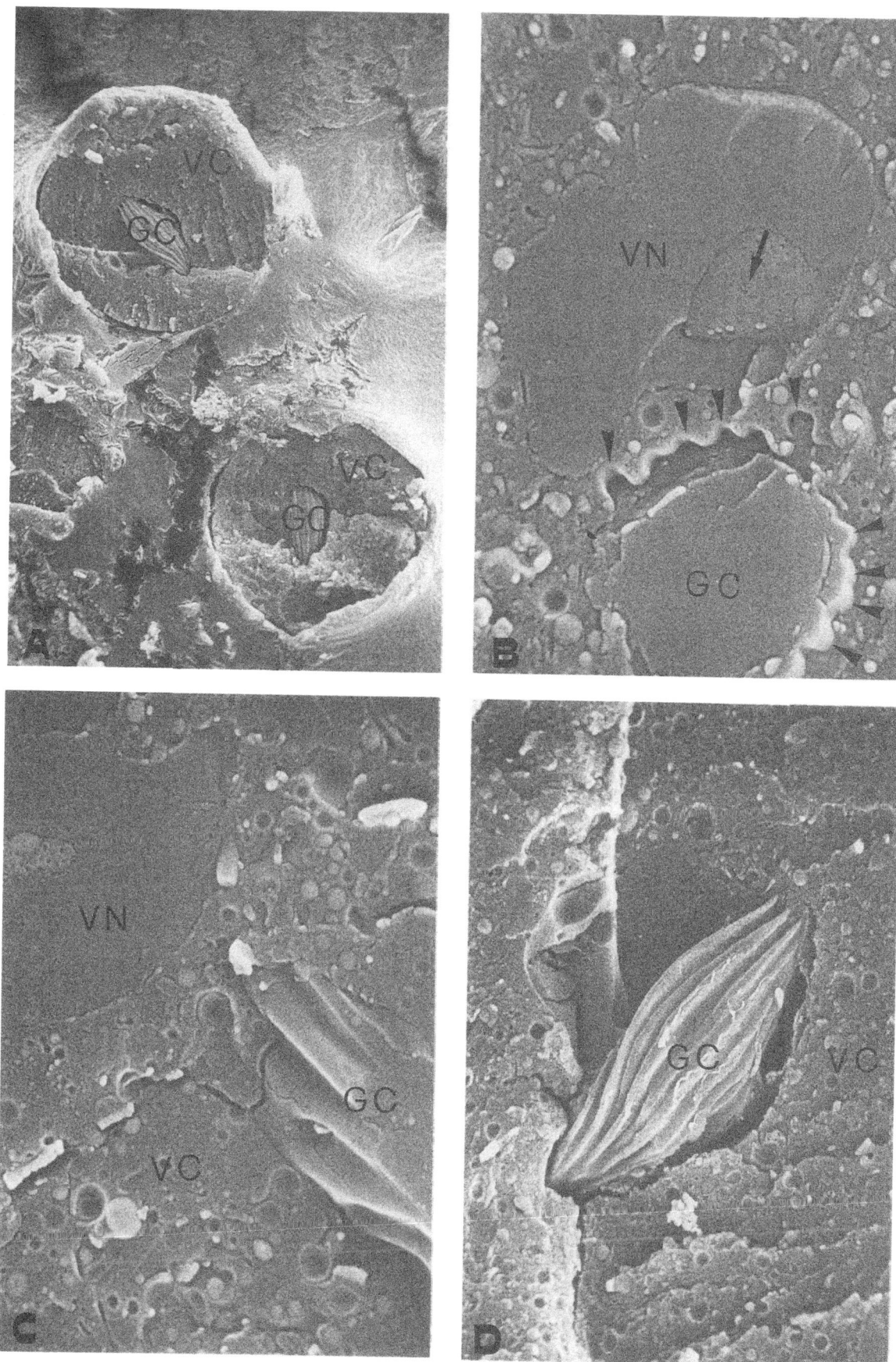

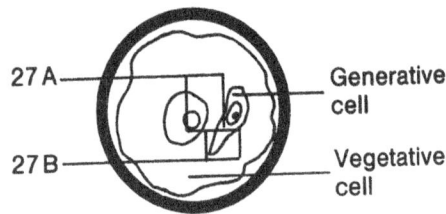

Plate 27A. Transverse section through the generative cell of *Nicotiana alata* Link & Otto, after conventional chemical fixation. The plasma membranes of the generative cell and the vegetative cell run fairly parallel to each other, but they are separated by a considerable space (arrowheads). In this space many components, of varying structure and morphology, can be seen. Some of these components have been interpreted as plasmodesmata, others have been thought to be constituents of a cell wall. x 29,000.

Plate 27B. Transverse section through the generative cell of *Nicotiana alata* Link & Otto, after rapid freezing and freeze substition. Although the cytoplasmic membranes are not clearly visible, the plasma membranes of the generative cell and the vegetative cell are distintly discernible (arrowheads). They now appear very straight and separated from each other by only a thin space. Moreover this space clearly is filled with electron-dense material, indicating the presence of a cell wall. The observation that this cell wall is only poorly fixed with conventional chemical fixation (plate 23A) suggests that it is not just a normal, solid, pecto-cellulosic cell wall. x 108,000.

(Plate 27A courtesy F Ciampolini, Siena; Plate 27B courtesy C Milanesi, Siena).

RER
V
M
GN
VN
M
VN
VC
GC
D
A

GN
GC
VC
B

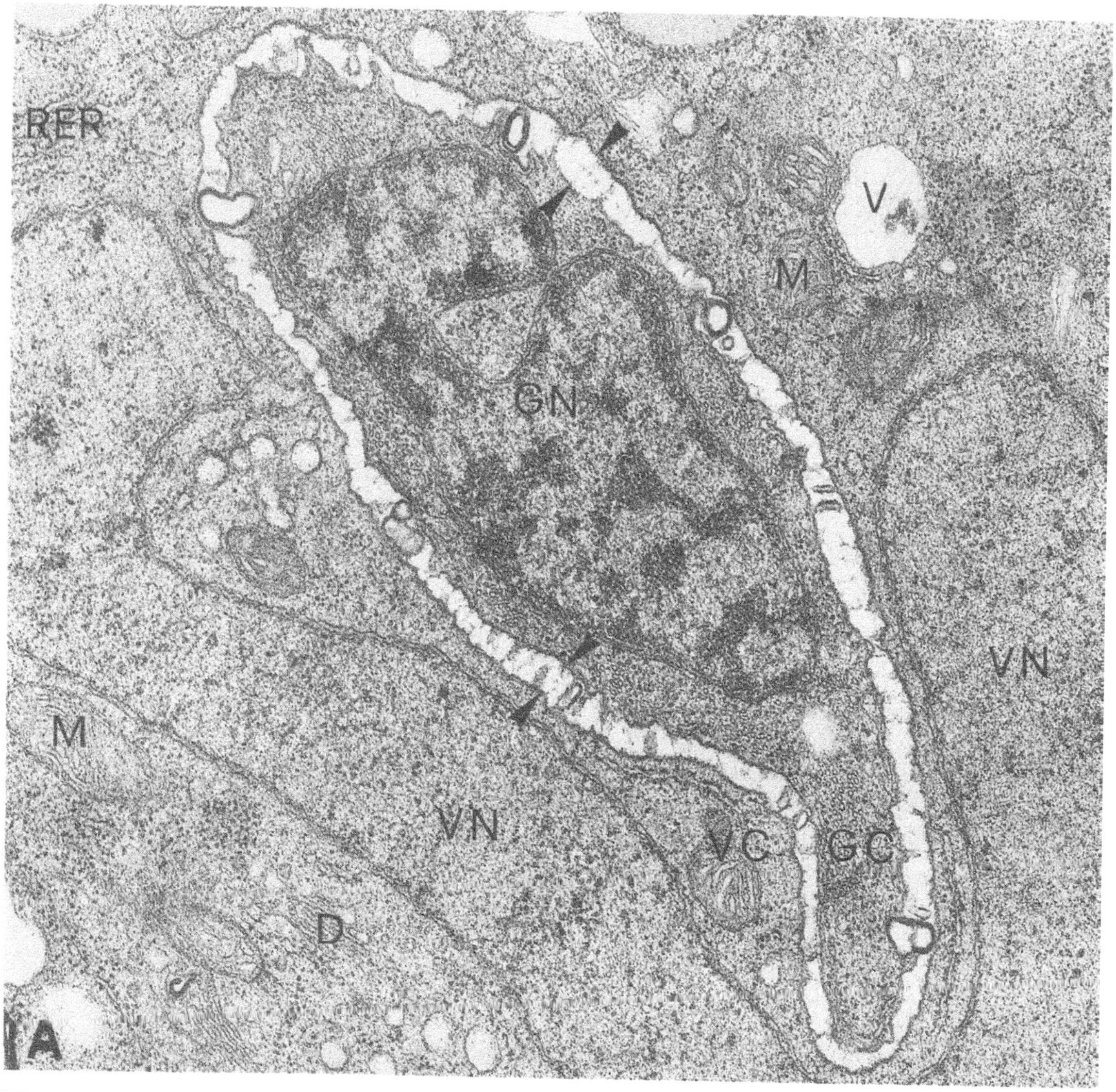

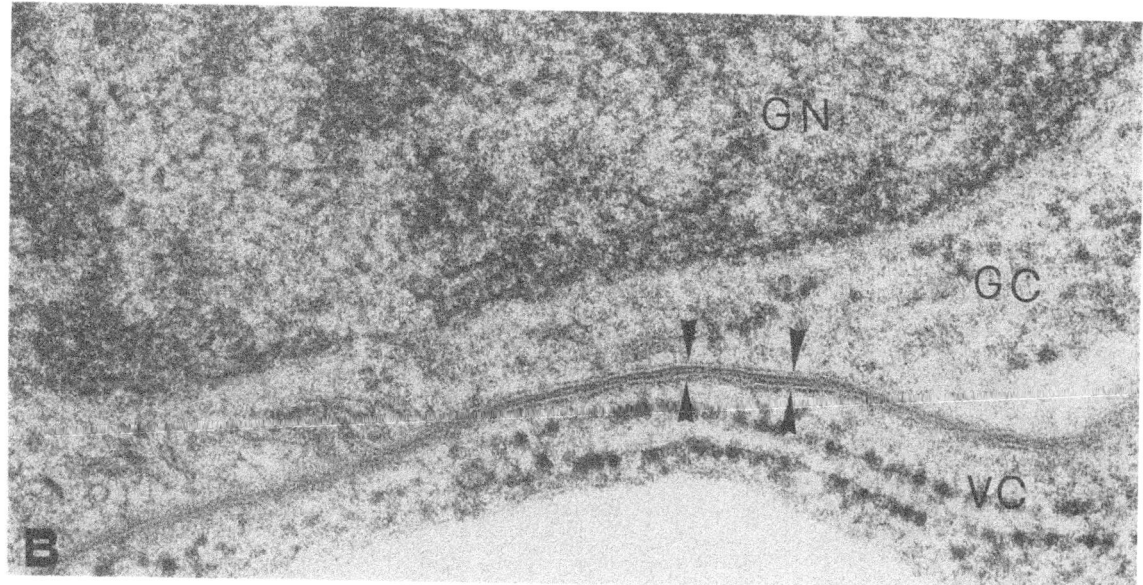

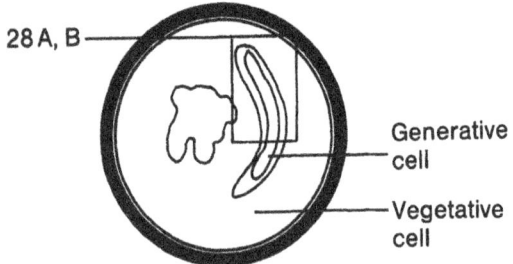

Plate 28A. Longitudinal section of the generative cell of a mature pollen grain of *Aloë ciliaris* Haw., prepared by conventional chemical fixation. It shows the longitudinally arranged bundles of microtubules. The generative cell wall shows the "swollen" appearance, typical for conventional chemical fixation with glutaraldehyde and osmium tetroxyde. x 40,000.

Plate 28B. Longitudinal section of the generative cell of a mature pollen grain of *Papaver rhoeas* L., prepared by rapid freezing and freeze substitution. It also shows the longitudinally arranged bundles of microtubules. The generative cell wall shows the electron-dense, straight appearance, typical for rapid freezing and freeze substitution. x 21,000.

(Plate 28A courtesy F Ciampolini, Siena; Plate 28B courtesy C Milanesi, Siena).

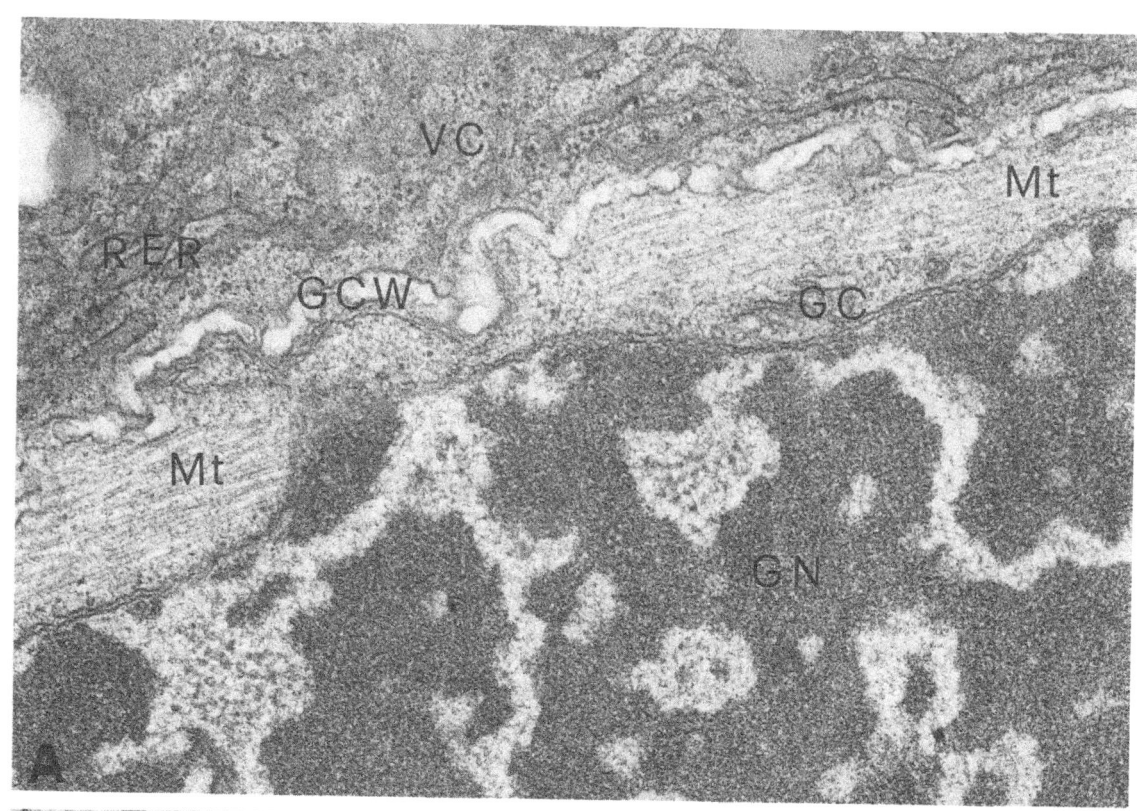

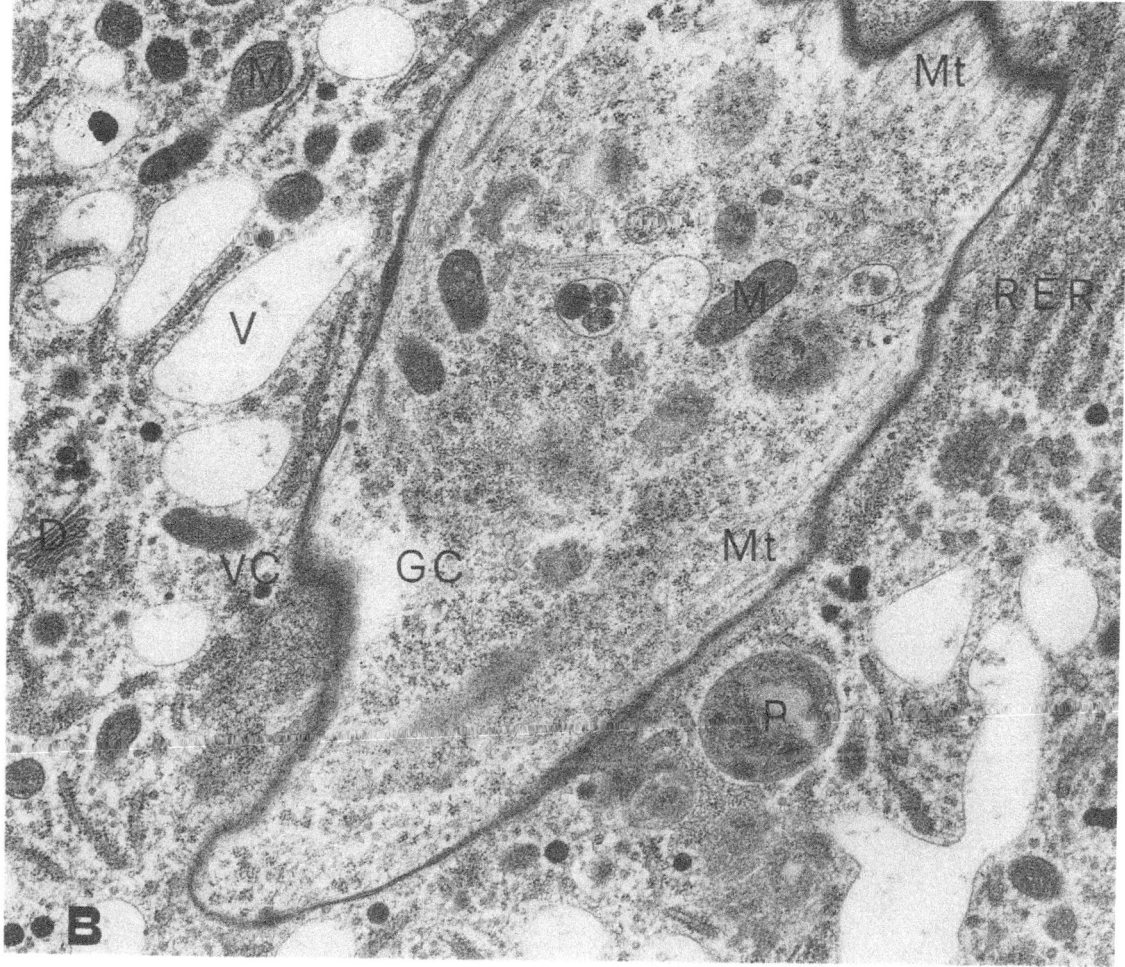

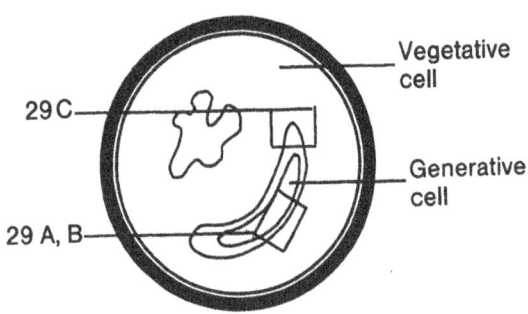

Plate 29A. Portion of the vegetative cell and generative cell of *Amaryllis bella-donna* L.. At the cytoplasmic face of the outer membrane of the generative cell groups of strip-shaped projections are present. The arrow indicates projections in transverse section, the double arrow points to projections in surface view. The projections are oriented perpendicular to the long axis of the generative cell, as indicated by the orientation of the microtubules (arrowheads). x 42,000.

Plate 29B. Enlarged portion of vegetative cell and generative cell of *Amaryllis bella-donna* L. showing the membrane projections in transverse section (arrows). x 75,000.

Plate 29C. Enlarged portion of vegetative cell and generative cell of *Amaryllis bella-donna* L., showing the membrane projections in surface view (arrowheads). x 70,000.

(Plates 29A-C reproduced by permission from M Cresti et al (1991) Ann Bot 68: 105-107).

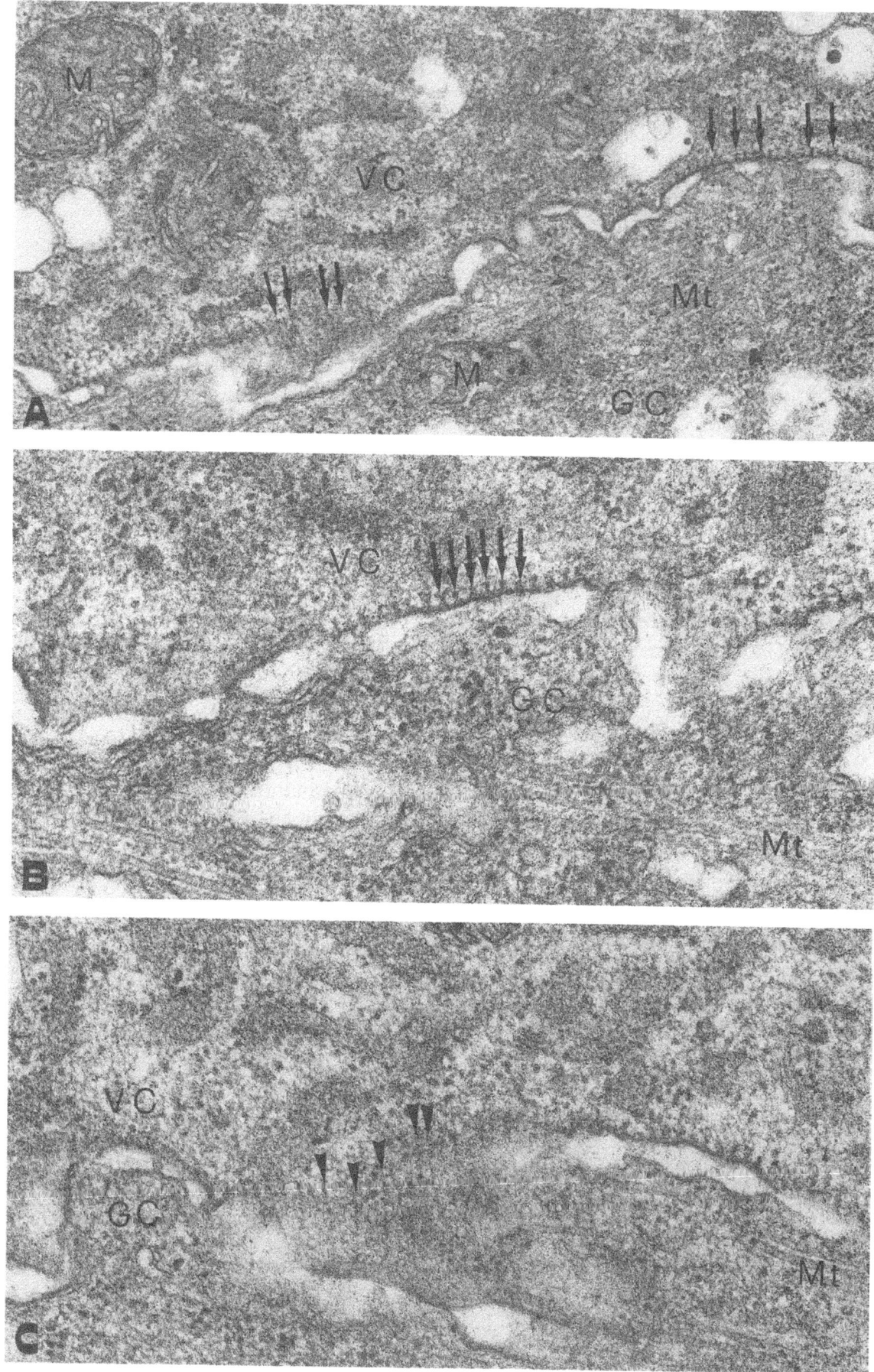

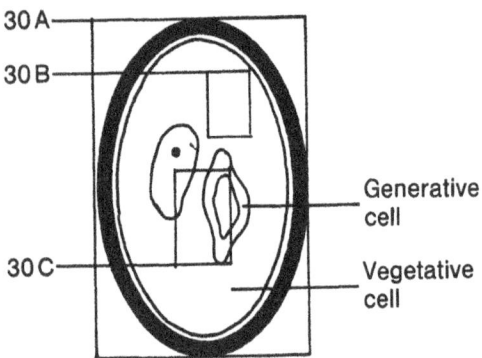

Plate 30A. Mature bicellular pollen grain of *Impatiens walleriana* Hook f. At the mature stage the cytoplasm of the vegetative cell is literally packed with Golgi-vesicles. x 2,800.

Plate 30B. Enlarged portion of the vegetative cell of the mature pollen grain of *Impatiens walleriana* Hook f. In between the numerous Golgi-vesicles many cisternae of smooth endoplasmic reticulum (arrows) are dispersed. Mitochondria are randomly distributed. x 29,000.

Plate 30C. Enlarged portion of the vegetative cell of the mature pollen grain of *Impatiens sultani* Hook f. The pollen has been freeze-fixed and freeze-substituted. After this type of fixation the Golgi-vesicles are strongly stained. The generative cell has long extension (arrows) which penetrate deeply into the highly lobed vegetative nucleus. x 18,000.

(Plates 30A,B reproduced by permission from Van Went J (1974) Fertilization in higher plants (HF Linskens ed) 81-88; Plate 30C courtesy C Milanesi, Siena).

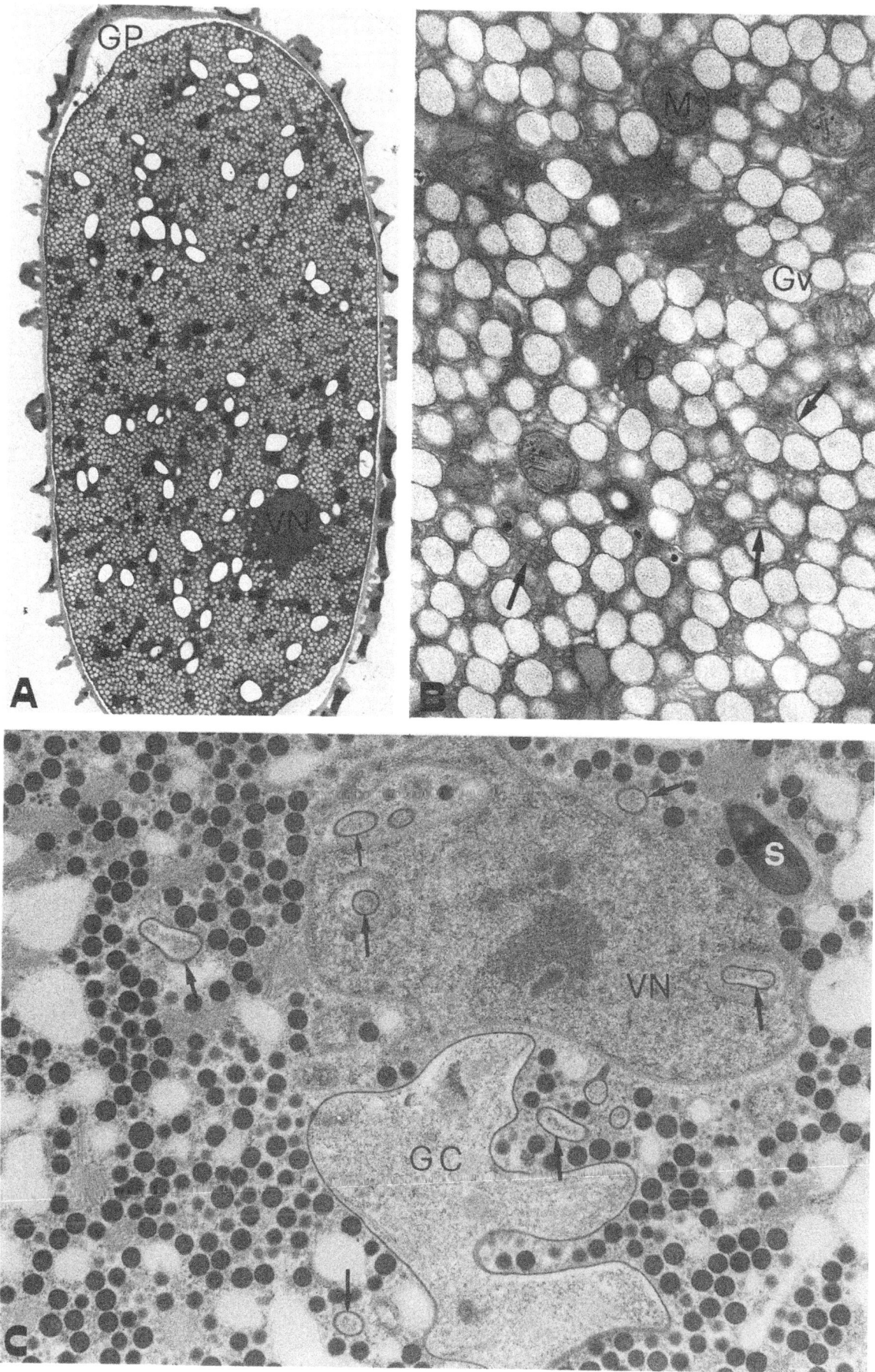

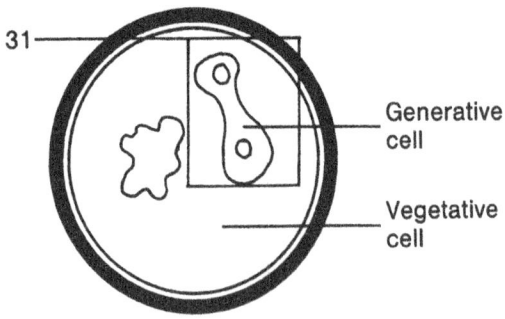

Generative
cell

Vegetative
cell

Plate 31. Division of the generative cell in the pollen grain of *Brassica napus* L..In *Brassica napus* the generative cell does form two sperm cells during the maturation of the pollen grain. Following mitosis the daughter nuclei become located in the peripheral regions of the elongated generative cell. At the equatorial plane a cell plate is formed centripetally (arrows).
x 25,800.

(Plate 31 reproduced by permission from M Charzynska et al (1989) Protoplasma 149: 1-4).

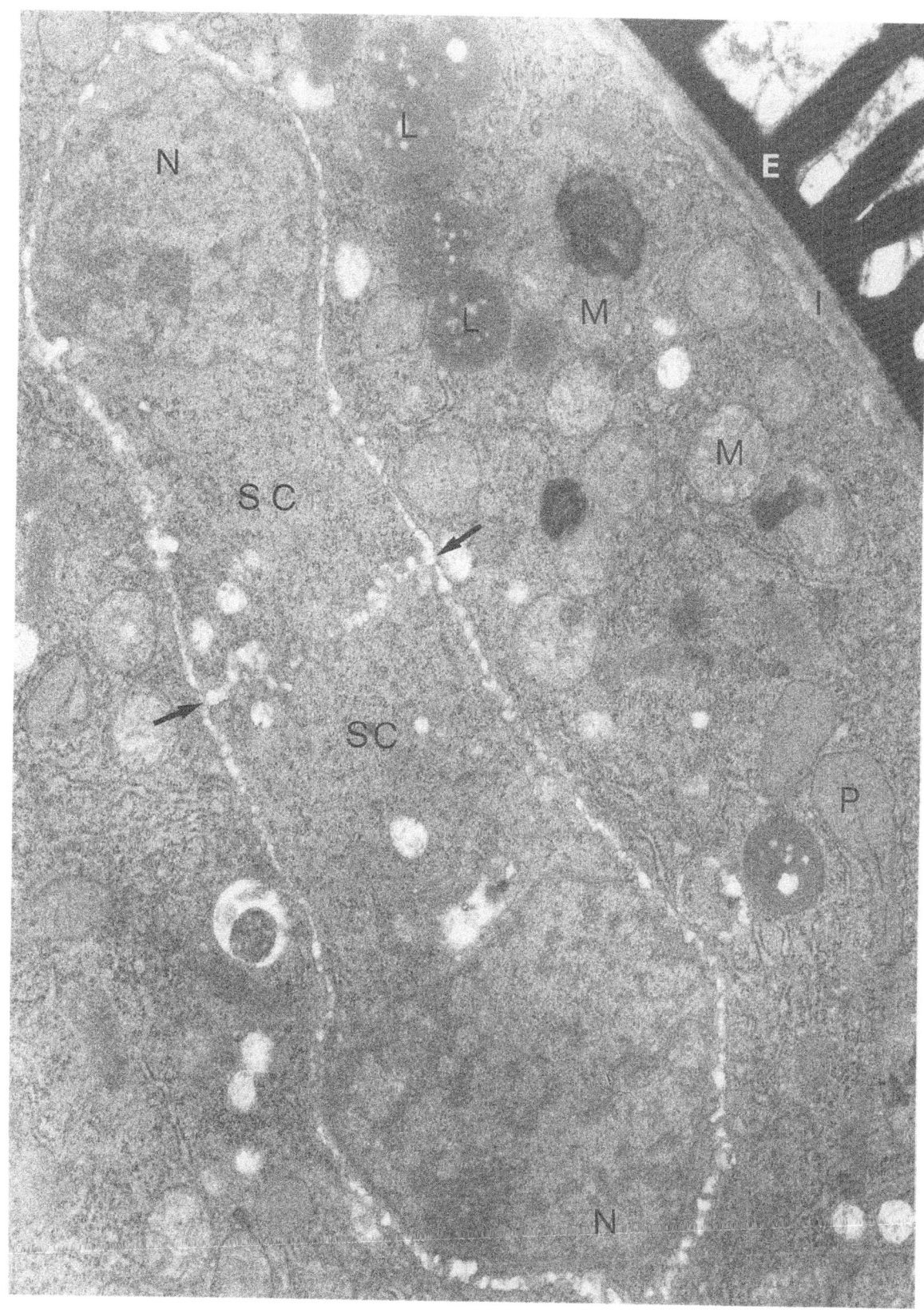

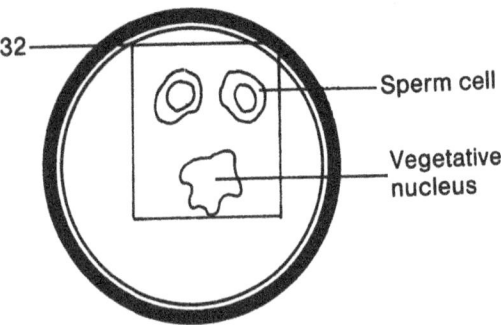

Plate 32. Portion of the mature tricellular pollen grain of *Hordeum vulgare* L. At the mature stage the pollen grain of *Hordeum* contains a large vacuole. In the presented portion of the grain the vegetative nucleus and the two, transversely sectioned sperm cells can be seen. The three elements usually are located close to each other, forming together the male germ unit. Note the presence of many channels in the intine layer of the pollen wall. x 10,300.

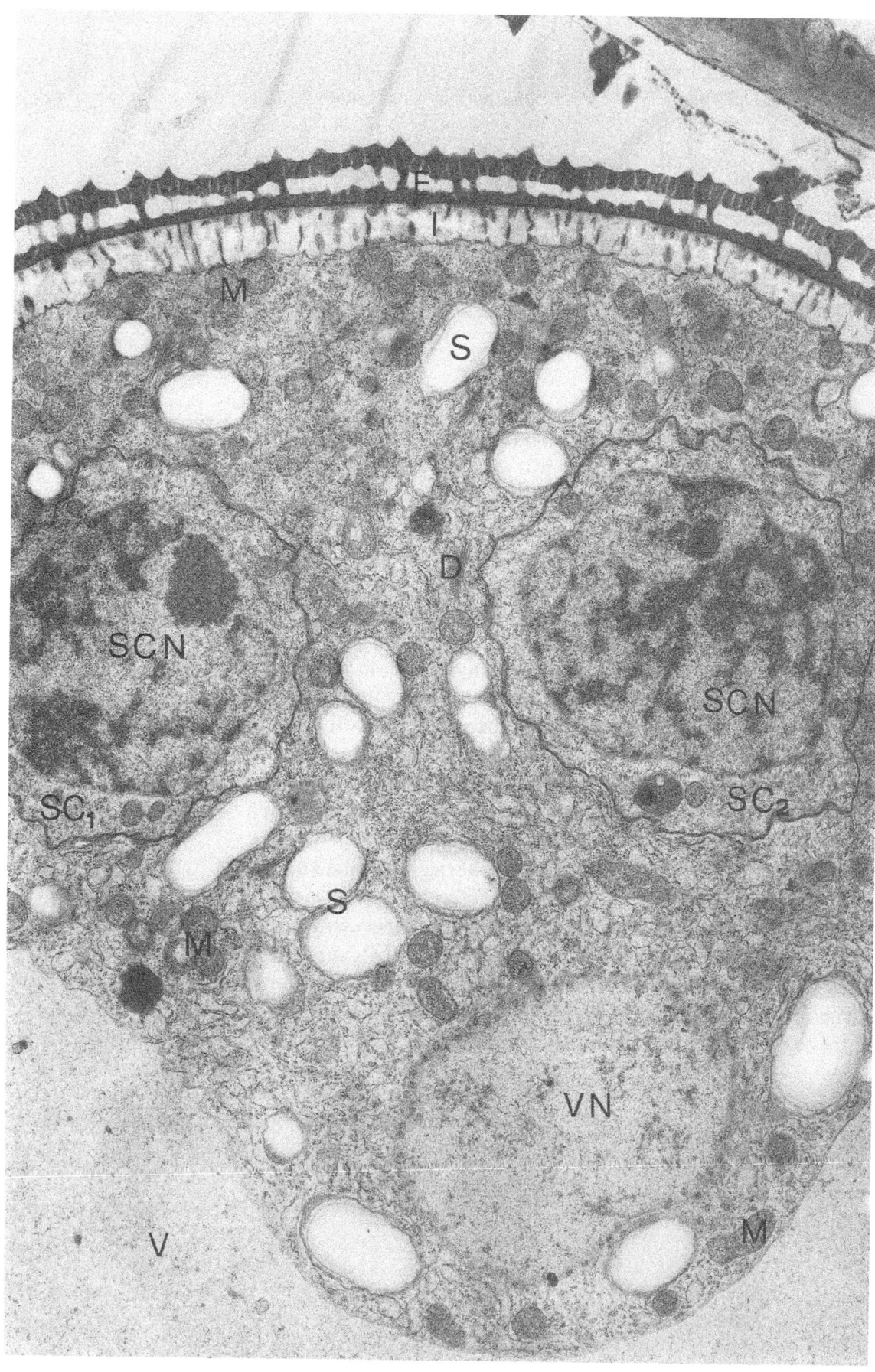

81

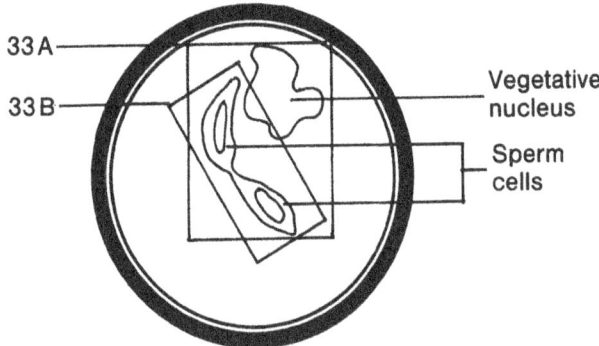

Plate 33A. Portion of a mature pollen grain of *Euphorbia dulcis* L., showing the two sperm cells. The two sperm cells are separated from each other by their respective plasma membranes. They remain, however, in close contact (arrows), since they are enclosed within one, continuous plasma membrane of the surrounding vegetative cell. x 19,500.

Plate 33 B-D. Isolated sperm cells of *Spinacia oleracea* L.. The sperm cells of *Spinacia oleracea* L. can be isolated from the pollen grains by mechanical squeezing and subsequent percoll density gradient centrifugation.
B. Directly after their release from the pollen grain, the two sperm cells still have their original shape and they are connected to each other. x 1,300.
C. Subsequently the sperm cells become spherical in shape and the paired configuration is lost. x 1,300
D. Ultrastructure of an isolated sperm cell of *Spinacia oleracea* L.. The isolated sperm cell has a spherical shape, and its nucleus is located in the centre of the cell. The cytoplasm is relatively simple in structure with a few mitochondria, dictyosomes and a limited amount of endoplasmic reticulum. x 24,500.

(Plate 33A courtesy M Murgia, Siena; Plate 33B, C reproduced by permission from Theunis C, Van Went J (1989) Sex Plant Reprod 2: 97-102; Plate 33D reproduced by permission from Theunis C (1990) Protoplasma 158: 176-181).

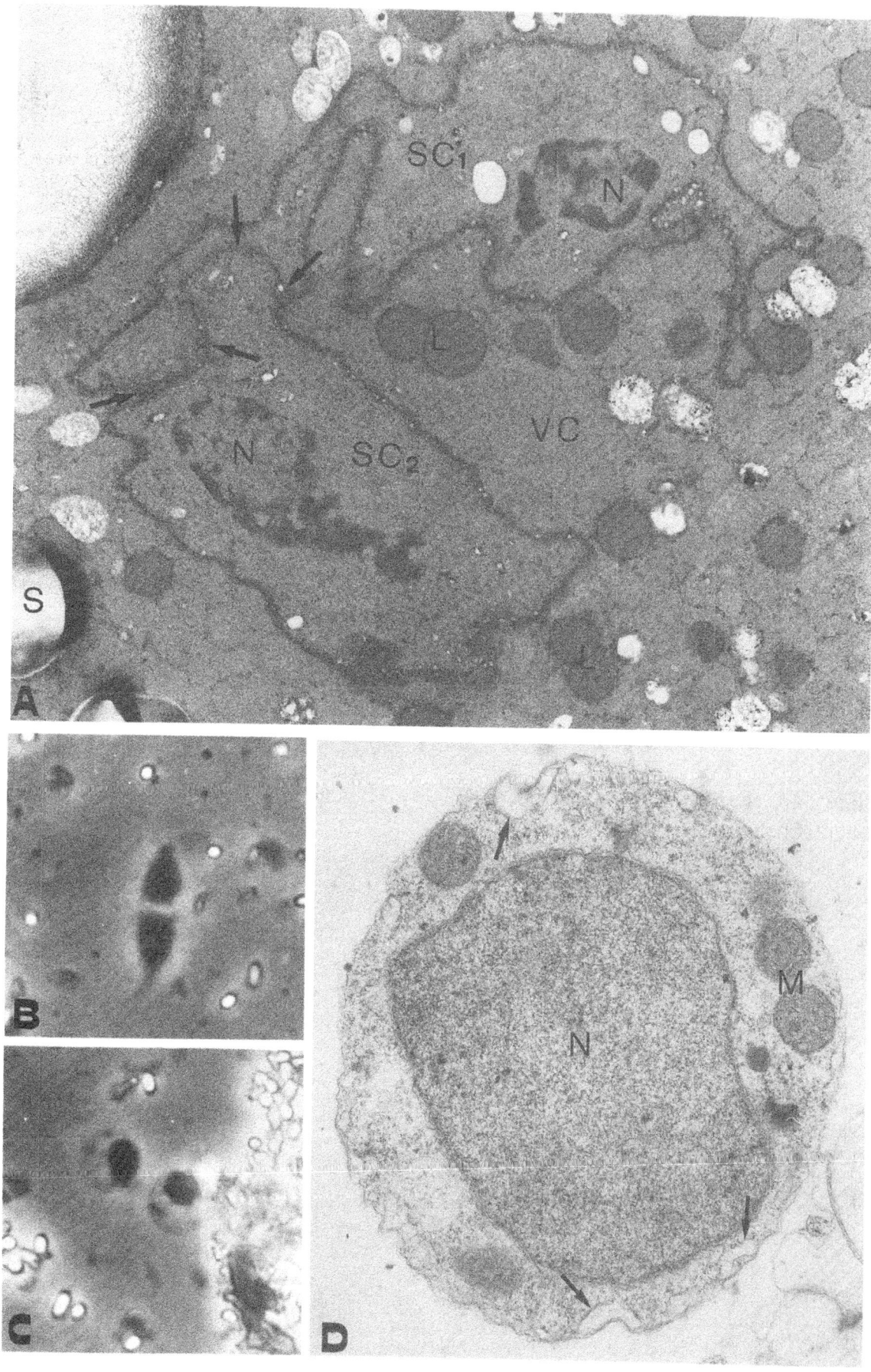

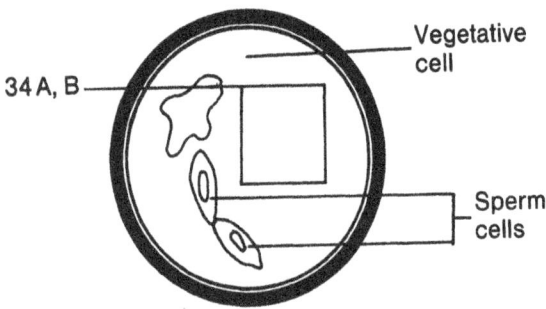

Plate 34A. Detailed view of the cytoplasm of the vegetative cell of the mature pollen grains of *Linaria vulgaris* Mill. The vegetative cell contains many mitochondria, simply structured plastids without starch, and an extensive system of rough endoplasmic reticulum of which the cisterns are dispersed throughout the entire cell. The cell is very rich in ribosomes. x 34,000.

Plate 34B. Detailed view of the cytoplasm of the vegetative cell of *Nicotiana alata* Link & Otto pollen. Whereas most organelles, as plastids, mitochondria and dictyosomes are randomly distributed, the rough endoplasmic reticulum forms large stacks. x 24,500.

(Plate 34A courtesy C Milanesi, Siena; Plate 34B courtesy F Ciampolini, Siena).

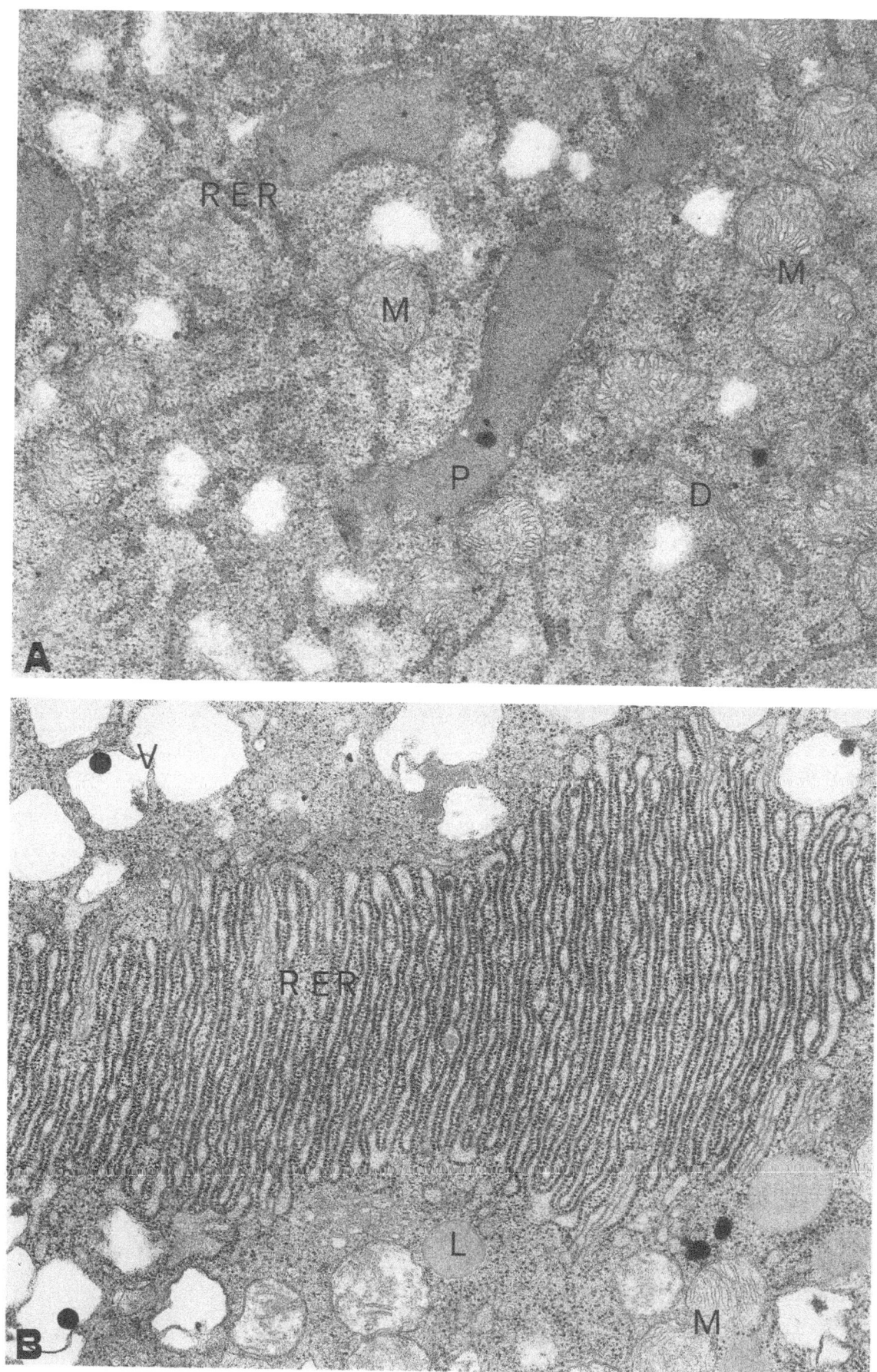

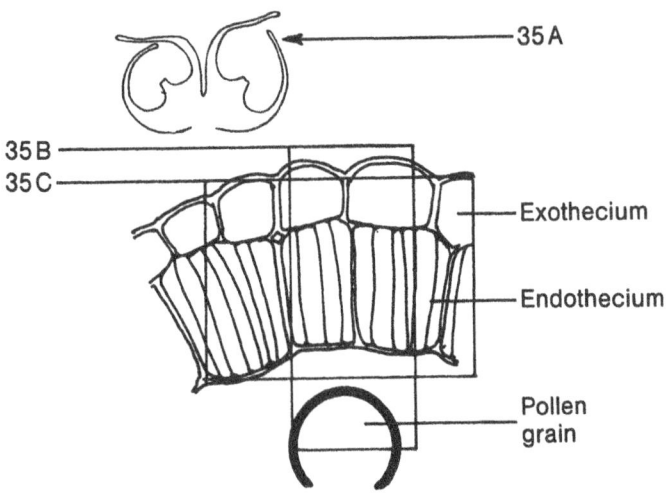

Plate 35A. Open anther of *Hibiscus rosa-sinensis* L.. After opening of the flower, the mature anthers rapidly loose water which results in the strong shrinkage of the anther wall. Because of a specially structured part of the wall, the stomium, and the specialized endothecium, shrinkage leads to the rupture of the wall and the exposure of the pollen grains. x 290.

Plate 35B. Structure of the anther wall of *Gasteria verrucosa* (Mill.) Duval at the time of anthesis of the flower. From the original locule wall only the outer layer, the exothecium and the endothecium are left. The middle layer and the tapetum are degenerated. From the tapetum cells only the deposits of sporopollenin, the orbicles remain (arrows). The thickenings of the tangential cell walls of the endothecium (arrowheads) are very prominent. Because of these thickenings the loss of water results in tangential shrinkage of the wall causing it to rupture at the stomium. x 2,200.

Plate 35C. Scanning electron micrograph of the endothecium of *Gasteria verrucosa* (Mill.) Duval, showing the orientation and distribution of the wall thickenings (arrowheads). x 1,100.

(Plate 35A courtesy F Ciampolini, Siena; Plate 35B reproduced by permission from Keijzer C (1987) New Phytol 105: 499-507; Plate 35C courtesy C Keijzer, Wageningen).

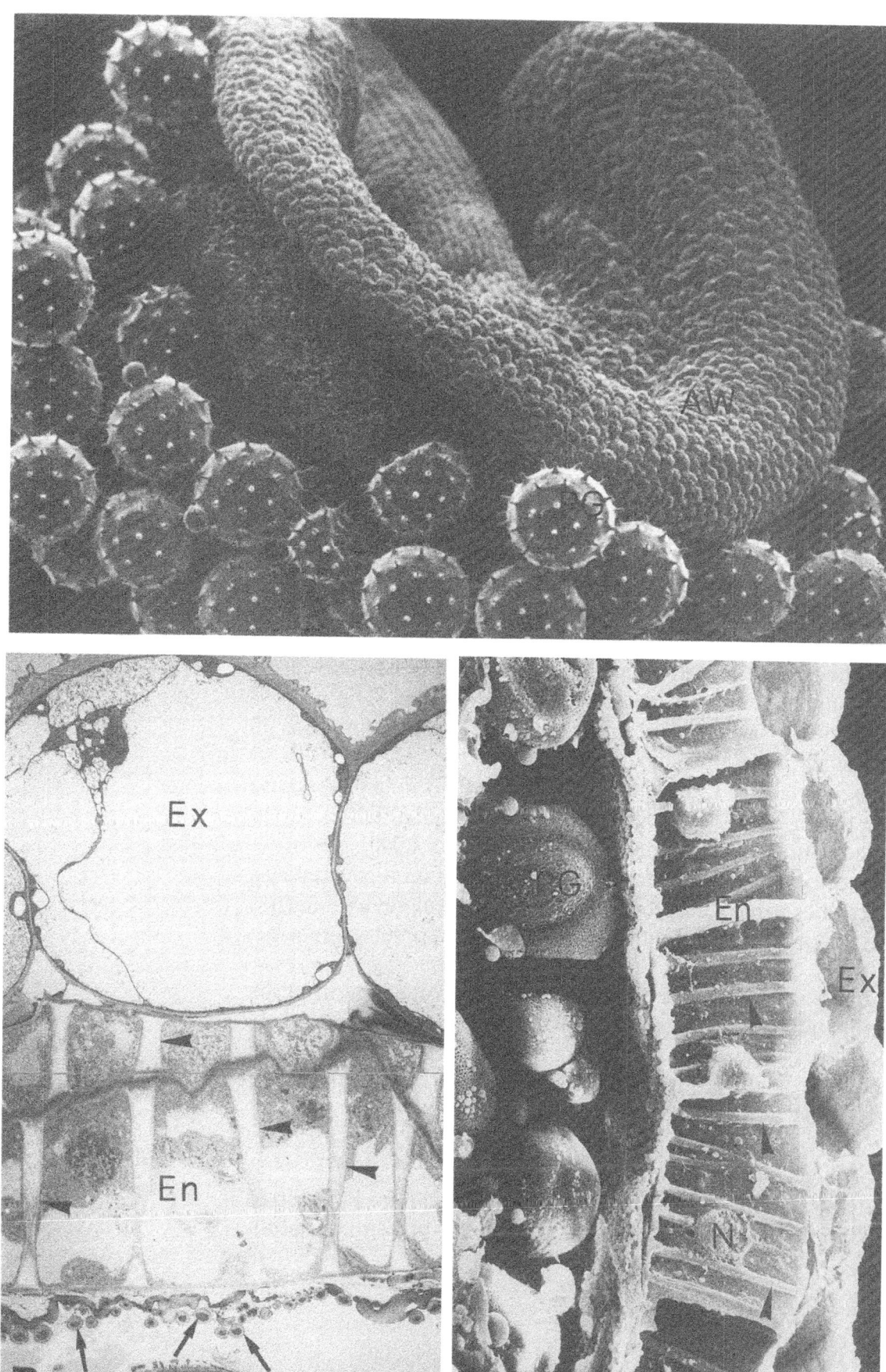

87

Plate 36. Different types of pollen grains.

Anemophilous (wind transported) grains.

A. *Chenopodium album* L. The spherical grain is pentaporate (with about 70 pores) (arrows), and apolar. Each circular pore has a diameter of about 1.3 μm. The exine is fairly smooth with minute spines (micro-echinate). x 2,000.

B. *Alopecurus myosuroides* Hudson. The spherical grain is monoporate (arrow), the only germination pore has an average diameter of 5.2 μm. The thin exine has only slight sculpturing elements (scabrate), so that the surface looks stippled. x 1,000.

C. *Picea abies* (L.) Karsten. The pollen grain is characterized by two large sacci, conspicuous, air-filled sacs. Between the sacs is a furrow (arrow). The exine texture of the cap is very fine and at the dorsal face of the body at least 5 μm thick. x 520.

D. *Acer negundo* L. The pollen grain is tricolpate (with 3 colpi) (arrows) and isopolar. The sculpturing elements of the exine are irregulary distributed (rugulate). x 2,000.

Entomophilous (insect transported) grains.

E. *Helianthus annuus* L. The nearly spheroidal isopolar pollen grain has apertures, longitudinal furrows with a central pore (tricolporate), of which only one is visible (arrow). The exine is covered with sharp, pointed spines (echinate) with perforations on the base of the spine. x 1,000.

F. *Galium mollugo* L. The pollen grain is stephanocolpate, because more than 3 furrows are visible (arrow), and isopolar. The exine has numerous small perforations and tiny spinules. x 4,000

G. *Brassica napus* L. The tricolpate, isopolar grain is covered by a rather thick exine which forms a network (reticulate) with narrow muri (ridges), in the lumina one can sometimes observe granulae. x 2,000

H. *Hedera helix* L. The pollen grain is tricolpate with three meridionally arranged germinal furrows (arrow), and isopolar. In polar view it appears triangular with rounded edges. The surface of the grain is reticulate, that is covered with a network of anastomosing ridges enclosing lumina. x 2,000.

(Plate 36 reproduced by permission from Ciampolini F, Cresti M (1981) Atlante dei principali pollini allergenici presenti in Italia).

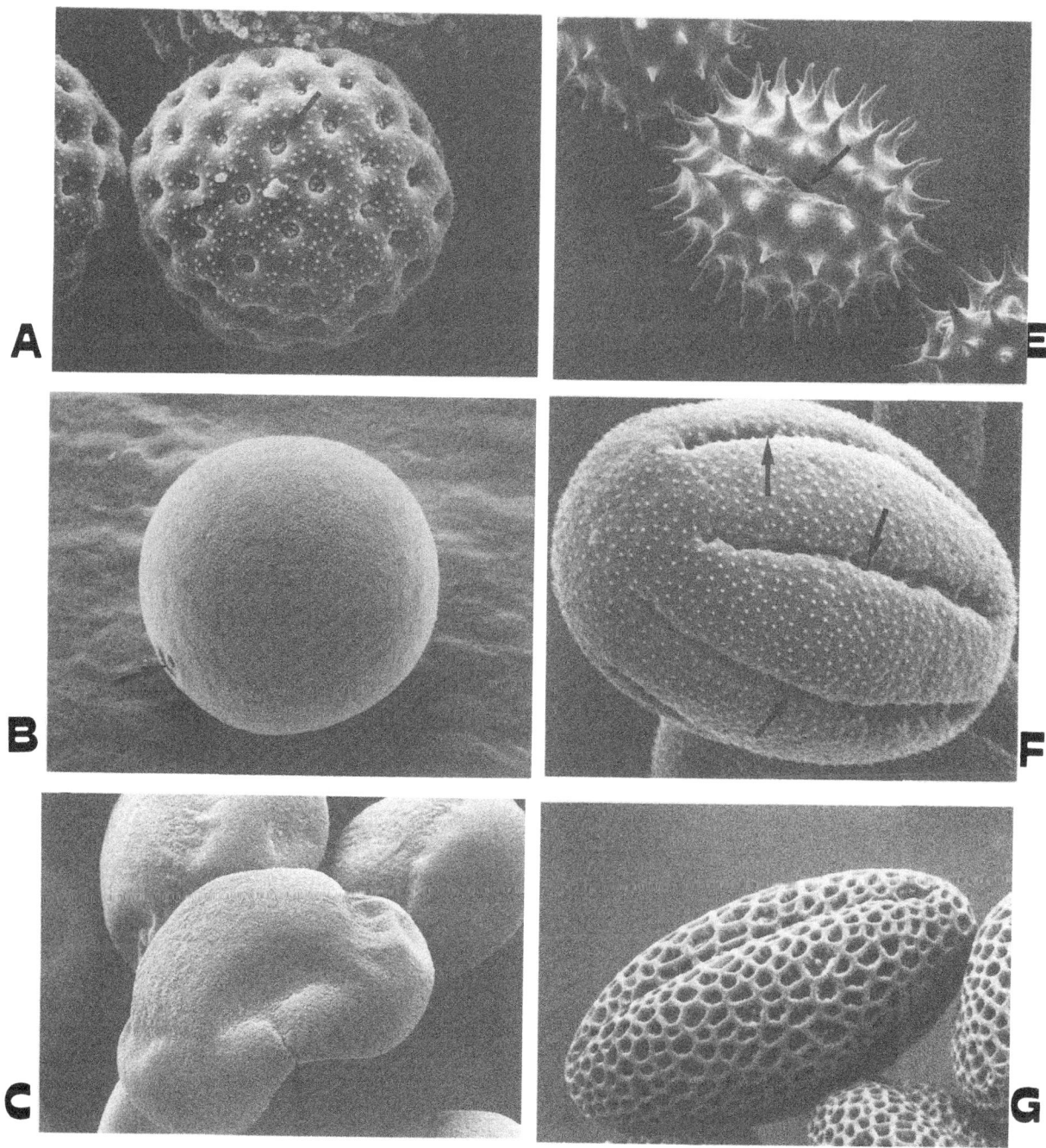

A

B

C

D

E

F

G

H

89

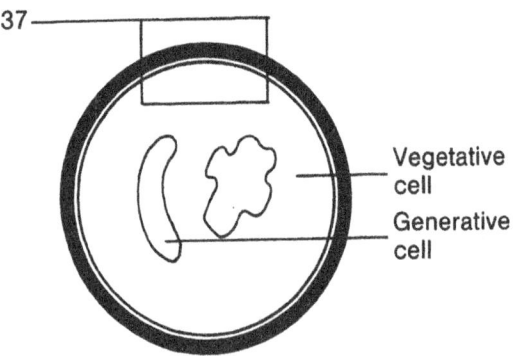

Plate 37A. *Bauhinia andrieuxii* Hemsl.: detail of mature exine of an acetolysed pollen grain showing dark staining ectexine and lighter staining endexine. x 4,000.

Plate 37B. *Bauhinia viridescens* Desv.: detail of mature exine of an acetolysed pollen grain showing dark staining ectexine consisting of a tectum with rounded supratectal processes overlying rod-like columellae and foot layer. The endexine is lighter staining. x 10,000.

Plate 37C. *Psoralea bituminosa* L.: detail of mature exine of an acetolysed pollen grain showing complex internal organisation of the ectexine which has an outer rod-like layer overlying a granular or spongy zone and a foot layer. x 4,000.

Plate 37D. *Indigofera polysphaera* Baker: detail of mature exine of an acetolysed pollen grain showing complex internal organisation of the ectexine which has a solid tectum overlying a spongy infratectal layer and a thick foot layer. x 4,000.

Plate 37E. *Cryptomeria japonica* D. Don: detail of the pollen wall of a mature, unacetolysed gymnosperm pollen grain. The ectexine is thinner and more lightly staining than the thick, distinctly lamellated endexine. x 20,000.

Plate 37F. *Ephedra disticha* L.: detail of the pollen wall of a mature unacetolysed gymnosperm pollen grain. The ectexine stains more lightly than the distinctly lamellated endexine and the intine is present beneath the exine. x 20,000.

Plate 37G. *Secale cereale* L.: detail of a mature pollen wall and adjacent tapetal membrane with Ubisch bodies which, like the exine, are acetolysis-resistant structures composed of sporopollenin. The thickened intine has numerous darkly stained radial channels. x 18,000.

(Plates 37 A-D courtesy I K Ferguson, Kew; Plates 37 E-F courtesy M H Kurmann, Kew; Plate 37G courtesy F Ciampolini, Siena).

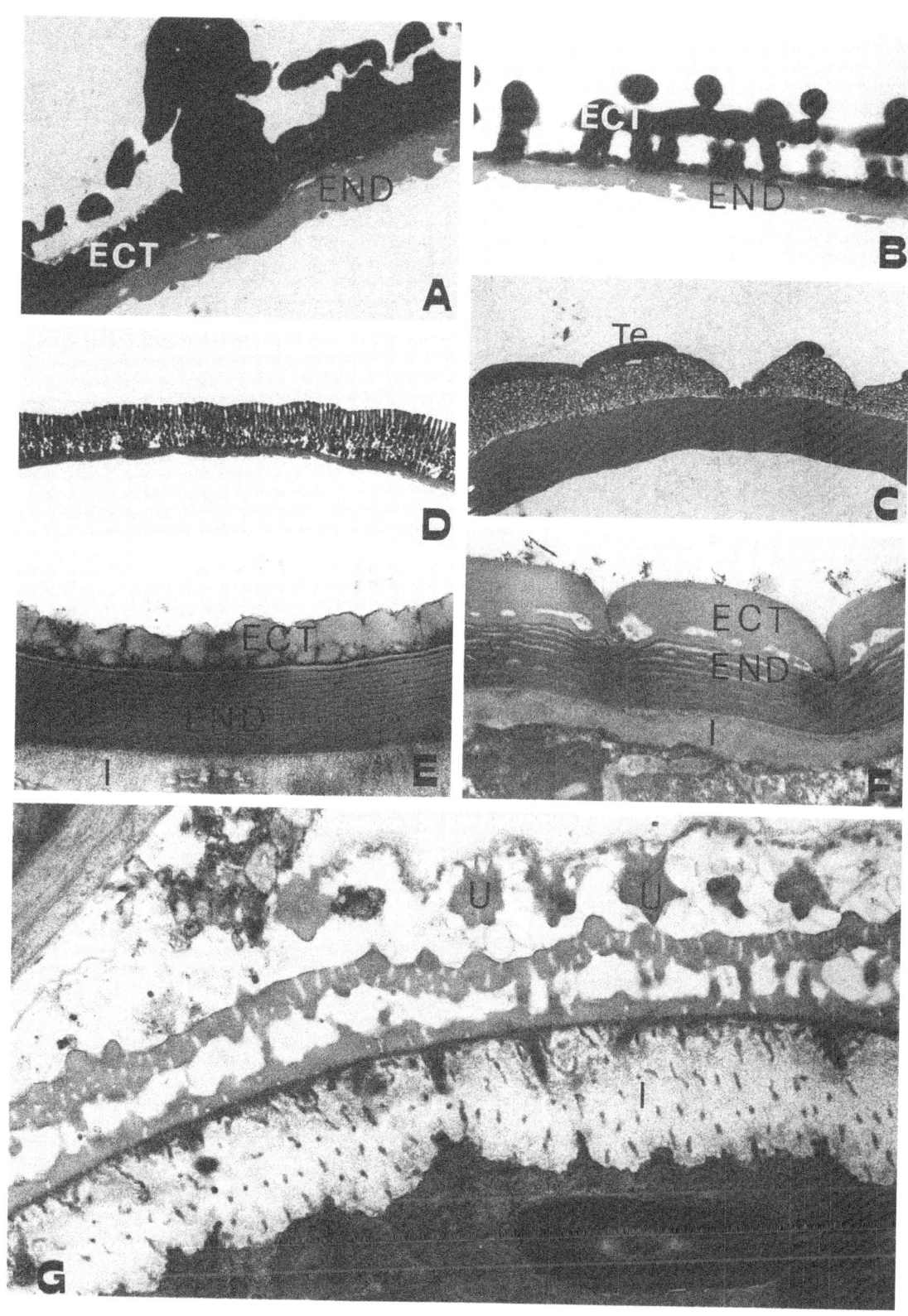

PART 2: PISTIL DEVELOPMENT

PISTIL DEVELOPMENT

Introduction

In flowering plants, the female reproductive organ, the pistil, is composed of three parts: the stigma, style and ovary. The stigma develops a receptive surface to which pollen grains adhere and which also provides the conditions for compatible grains to germinate. The stigma is situated at the apex of the generally elongated style, which places it in a favourable position to receive the pollen grains, while the style provides the pathway for the growing pollen tube towards the ovary. Inside the ovary, ovules containing the megasporangia develop at specific regions of the placenta. The young ovule is composed of a bulbous, multicellular tissue, the nucellus, which is positioned on top of a short stem, the funicle. The developing nucellus becomes enveloped by one or two integuments leaving only a narrow opening, the micropyle, at the apex through which the pollen tube can pass. The ovule can become curved or inverted during development. In the most common arrangement, the anatropous condition, the ovule is completely inverted, so that the micropyle becomes positioned close to the placenta.

At an early stage of ovule development, one cell of the nucellus develops into the megaspore mother cell. The megaspore mother cell is conspicuous because of its much larger size, denser cytoplasmic contents and more prominent nucleus. It divides meiotically and, generally, each nuclear division is followed directly by cytokinesis, resulting in a linear array of four haploid megaspores. In some taxa, however, cytokinesis occurs only after the first meiotic division, leading to the formation of binucleate (coenocytic) megaspores. Cytokinesis can also be completely absent, with the megaspore mother cell giving rise to a four-nucleate (coenocytic) megaspore.

As a rule, only one of the four resulting megaspores persists, while the others degenerate. The persisting, functional megaspore, which thus can be mono-, bi- or tetranucleate, develops into the embryo sac, the female gametophyte.

During this megasporogenesis, callose is deposited in the walls of the megaspore mother cell and megaspores, similar to the callose deposition during microsporogenesis. Callose always appears in the early meiotic prophase. As a result the cells become temporarily isolated from the surrounding sporophytic tissue, enabling an independent course of development to occur, accompanied by the shift from sporophytic to gametophytic gene expression. After completion of megasporogenesis the callose disappears from the wall of the functional megaspore, whereas it frequently remains present in the walls of the megaspores which will degenerate.

In many species an uneven distribution of cell organelles is evident in the megaspore mother cell prior to cytokinesis. Additionally, in such cases, cell plate formation during cytokinesis occurs in a non-equatorial position, resulting in a larger cell which will become the functional megaspore, and a smaller cell which will degenerate.

The pattern of callose deposition and breakdown, the uneven distribution of cell organelles, and the unequal positioning of cell plates all clearly demonstrate the presence of a strong polarity in the developing megaspore mother cell.

As mentioned above, in the majority of angiosperms four haploid megaspores are formed, of which the three micropylar megaspores degenerate while the chalazal one develops into the embryo sac.

Embryo sac formation or megagametogenesis takes place in three stages. The first is the formation of a coenocytic embryo sac with eight haploid nuclei. The second is the cellularization of the embryo sac, followed by a third stage during which cell differentiation takes place. After the first mitotic division of the megaspore nucleus, the cell starts to enlarge, and this is accompanied by the formation of a large, central vacuole. The two daughter nuclei each take a peripheral position and remain in this position during their subsequent divisions. The resulting coenocytic, eight-nucleate embryo sac contains four nuclei in the micropylar region and four nuclei in the chalazal region, while the central region is highly vacuolate. The subsequent division into cells through the formation of cell plates results in three, small haploid cells in the micropylar region of the embryo sac, three small haploid cells in the chalazal region, and a large central cell containing two haploid nuclei. At this stage of development the cytoplasmic composition of the cells is basically similar and they are separated from each other by thin pecto-cellulosic cell walls. This similarity is, however, soon lost during the subsequent phase of differentiation.

The three cells at the micropylar pole of the embryo sac differentiate into the egg apparatus, consisting of two synergids and one egg cell. During differentiation the cells enlarge dramatically by local expansion of the walls that separate them from the central cell. The mature, pear shaped cells remain attached to the embryosac wall near the micropyle, hanging like droplets into the central cell. Each cell type, synergid and egg cell, develops a characteristic ultrastructure and organization.

The synergids are involved in the transfer of the male gametes from the pollen tube to the female gametes, the egg cell and the central cell. The micropylar half of the synergid is very rich in cytoplasm, while the chalazal half becomes highly vacuolated. The nucleus is positioned in the centre of the synergid, just above the vacuole. The portion of the synergid cell wall, bordering the micropyle and the adjacent part of the wall in between the two synergids, develops into the filiform apparatus by local and irregular thickening. At maturity the filiform apparatus consists of a mass of wall projections extending deep into the cytoplasm, resulting in an extensive plasma membrane surface area in that region. The filiform apparatus with its extensive plasma membrane surface is thought to be involved in the secretion of metabolic products of the synergids into the micropyle. Such metabolic products might be involved in a chemotropic gradient, capable of directing pollen tube growth towards the embryo sac. Certainly, the distribution and high number of organelles in the synergid cytoplasm indicate a high metabolic activity in the micropylar half of the cell. Synergid cytoplasm usually contains large numbers of mitochondria and dictyosomes, an extensive endoplasmic reticulum and abundant ribosomes.

From the filiform apparatus toward the base of the synergids the cell wall gradually decreases in thickness, so that the chalazal halves of the cells are bordered by their plasma membranes only.

Frequently, one of the synergids starts to degenerate prior to the arrival of the pollen tube, while the other retains its original condition. Degeneration is accompanied by the deposition of osmiophilic materials onto the membranes of the mitochondria, and the disappearance of the plasma membrane. The degenerated cytoplasm of this synergid then penetrates in between the other cells of the embryo sac. Invariably, it is into the degenerated synergid that the pollen tube injects its contents, including the sperm cells, during fertilization.

The egg cell, like the synergids, also shows a distinct polarity. However, the micropylar two-thirds of the cell contain a large vacuole, while most of the cytoplasm, with the nucleus, is located in the chalazal third of the cell. Usually, the egg cell is only partially surrounded by a cell wall. The egg cell wall is thickest at the micropylar pole of the cell, and from there gradually reduces in thickness towards the chalazal region where only a plasma membrane is present. Cell wall thickenings and extensions, as frequently observed in the other cells of the embryo sac, are not present in the egg cell. The ultrastructure of the egg cell cytoplasm varies considerably among species. In general, however, the egg cell contains less organelles than the other embryo sac cells, and appears to be relatively inactive metabolically .

The central cell occupies the largest portion of the embryo sac. It is highly vacuolate, and most of its cytoplasm is accumulated near the egg apparatus. The two polar nuclei, or their fusion product, the fusion nucleus, are also positioned here. At its micropylar pole, where it borders the egg apparatus, the central cell lacks a normal cell wall. Frequently, the micropylar portion of the lateral wall of the central cell shows many finger-like projections, which strongly resemble the filiform apparatus of the synergids. In most species the central cell contains cytoplasm with complex organization, rich in organelles. But, as with the egg cells, there is considerable diversity.

The three cells which are formed at the chalazal side of the embryo sac develop into antipodal cells. The antipodal cells show great diversity among angiosperms. In most taxa the antipodal cells degenerate before or during the maturation of the embryo sac. In other species, however, they persist even during embryo and endosperm formation, although their condition can be highly variable. Particularly in grasses, the antipodal cells are very conspicuous, sometimes proliferating into a multicellular tissue consisting of up to 100 cells. This great diversity makes it rather difficult to generalize concerning the structure and functions of antipodal cells.

The above description applies to the most common type of embryo sac development and structure, the Polygonum type. It starts from one uninucleate megaspore and results in a seven- celled, eight nucleate embryo sac. There are a large number of other developmental types, which can differ in many aspects, such as the number of nuclei in the functional megaspore, the number of mitotic divisions, the positioning of the nuclei, fusion of nuclei, and the number and position of cells that are formed. These differences mainly result in other ploidy levels of the nuclei in the central cell and antipodal cells, whereas in general the egg apparatus comprises three haploid cells, as in the Polygonum type.

As the embryo sac develops and changes, so does the surrounding nucellar tissue from which it is derived. In general, two developmental pathways are recognised, resulting in the so-called crassinucellate and tenuinucellate ovules. In the crassinucellate type, the nucellus tissue expands by cell division during embryo sac development, and the mature embryo sac is surrounded by a massive amount of nucellus. In the tenuinucellate type, the nucellus cells bordering the embryo sac subsequently degenerate and disappear during embryo sac development, and the mature embryo sac is encapsulated by the integuments only. Frequently, in the tenuinucellate type, the integument epidermis bordering the embryo sac develops a specific structure, the endothelium or integumentary tapetum. The cells of the endothelium are cytoplasm rich with prominent nuclei and become radially stretched.

Together with the developing ovary, the style and stigma also develop characteristic organisation and structure related to their reproductive functions.

The morphology of the mature stigma shows great diversity. Two principal categories of stigmas are recognised, those with copious fluid secretion - wet stigmas - and those with limited surface secretion - dry stigmas.

Wet stigmas either have a receptive surface with small to medium size papillae, or a nonpapillate surface. The stigma cells are glandular in nature and at maturity they produce and secrete the exudate which completely covers the surface. Sometimes, additional components are added to the exudate, as a result of degeneration of the stigmatic cells. The composition of the exudate shows considerable diversity among species. It usually contains lipophilic substances and polysaccharides, but many other components can be found as well.

Dry stigmas can also be subdivided into stigmas with non-papillate surfaces, and those with papillae. The papillae can be either unicellular or multicellular. The dry stigma surface has a cuticle covered with a pellicle. The latter is an extracellular, hydrated protein film, originating from the stigmatic cells.

Like the stigmas, styles also show great diversity among species. In general, two main types are distinguished. The open or hollow style contains a canal lined with a glandular epidermis, through which the pollen tubes grow toward the ovary. The solid style contains a solid core of transmitting tissue, which provides the pathway for the pollen tubes.

Transmitting tissue is composed of elongated cells, of which the longitudinal cell walls either have a thickened primary wall layer, or a thickened middle lamella, or: an additional intercellular substance is deposited between cells. It is through these extracellular compartments that the pollen tubes grow downwards to the ovary. The transverse walls between the cells generally are very thin and have large numbers of plasmodesmata.

The stylar canal of the hollow style becomes filled with liquid exudate during maturation. The materials are produced and secreted by the glandular canal cells. These cells are rich in organelles and frequently the cell walls bordering the canal show extensions, similar to those of the walls of transfer cells. The liquid exudate in the style provides the substratum for the growing pollen tubes.

Recommended literature

Cresti M, Ciampolini F, Van Went J L, Wilms H J (1982) Ultrastructure and histochemistry of *Citrus limon* (L.) stigma. Planta 156: 1-9

Erdelská O, Ciamporová M, Lux A, Pretová A, Tupy J (eds) (1983) Fertilization and embryogenesis in ovulated plants. Veda, Bratislava

Iwanami Y, Sasakuma T, Yamada Y (1988) Pollen: illustrations and scanning electronmicrographs. Springer, Berlin-Heidelberg-New York-Tokyo

Johansen D A (1950) Plant embryology, embryogeny of the spermatophyta. Chronica Botanica, Waltham, Massachusetts

Maheshwari P (1950) An introduction to the embryology of angiosperms. Mc Graw-Hill, New York.

Willemse M T M, Van Went J L (eds) (1985) Sexual reproduction in seed plants, ferns and mosses. Pudoc, Wageningen.

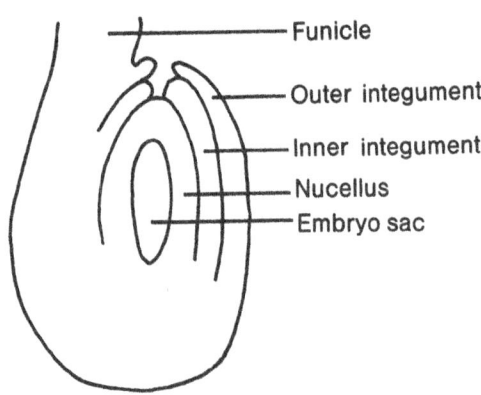

Plate 38A. Ovular primordia of *Passiflora racemosa* Brot., showing the initiation of the inner (arrow) and outer (arrowhead) integument. x 200.

Plate 38B. Young ovules of *Passiflora vespertilio* L.. Anatropous curvature of the ovule has started, and the inner (arrow) and outer integument (arrowhead) are overgrowing the nucellus (double arrow). The ovule is connected to the placenta through the funicle. x 200.

Plate 38C. Mature ovules of *Agrostemma gracilis* Boiss. The inner and outer integument have completely overgrown the nucellus, leaving only a small opening, the micropyle (arrow) at the ovule apex. Curvature of the ovule is only partially. x 200.

Plate 38D. Mature anatropous ovules of *Passiflora vespertilio* L.. Curvature of the ovule has resulted in complete inversion, and the micropyle (arrow) is positioned near the funicle and the placenta. x 150.

(Plates 38A-D courtesy F Bouwman, Amsterdam).

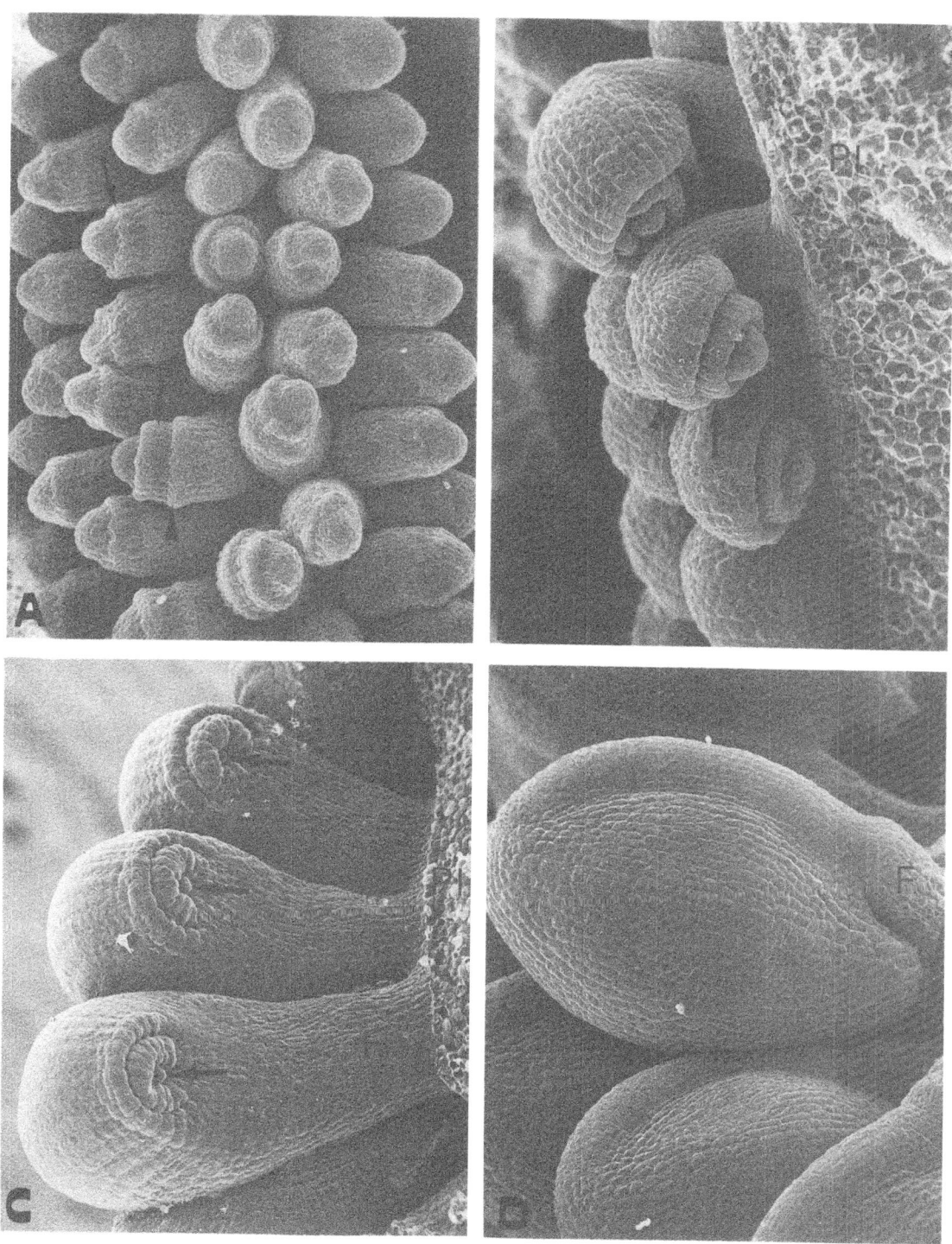

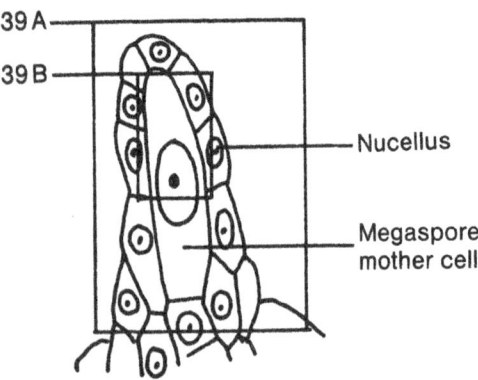

Plate 39A. Young ovule of *Impatiens walleriana* Hook f., with a megaspore mother cell at early prophase. The megaspore mother cell develops from a subepidermal nucellar cell which strongly enlarges. At this stage of ovule development integuments have not yet been formed. Note the large volume of the megaspore nucleus, as compared to those of the nucellus cells. x 3,100.

Plate 39B. Enlarged portion of the megaspore mother cell of *Impatiens walleriana* Hook f. at early prophase. The micropylar pole is at the left side of the photograph. The nuclear envelope and the nucleolus are still intact. At this stage of development the nucleus is positioned in the centre of the cell. The cytoplasmic constituents are randomly distributed. x 15,000

(Plates 39A courtesy M de Boer-de Jeu, Wageningen).

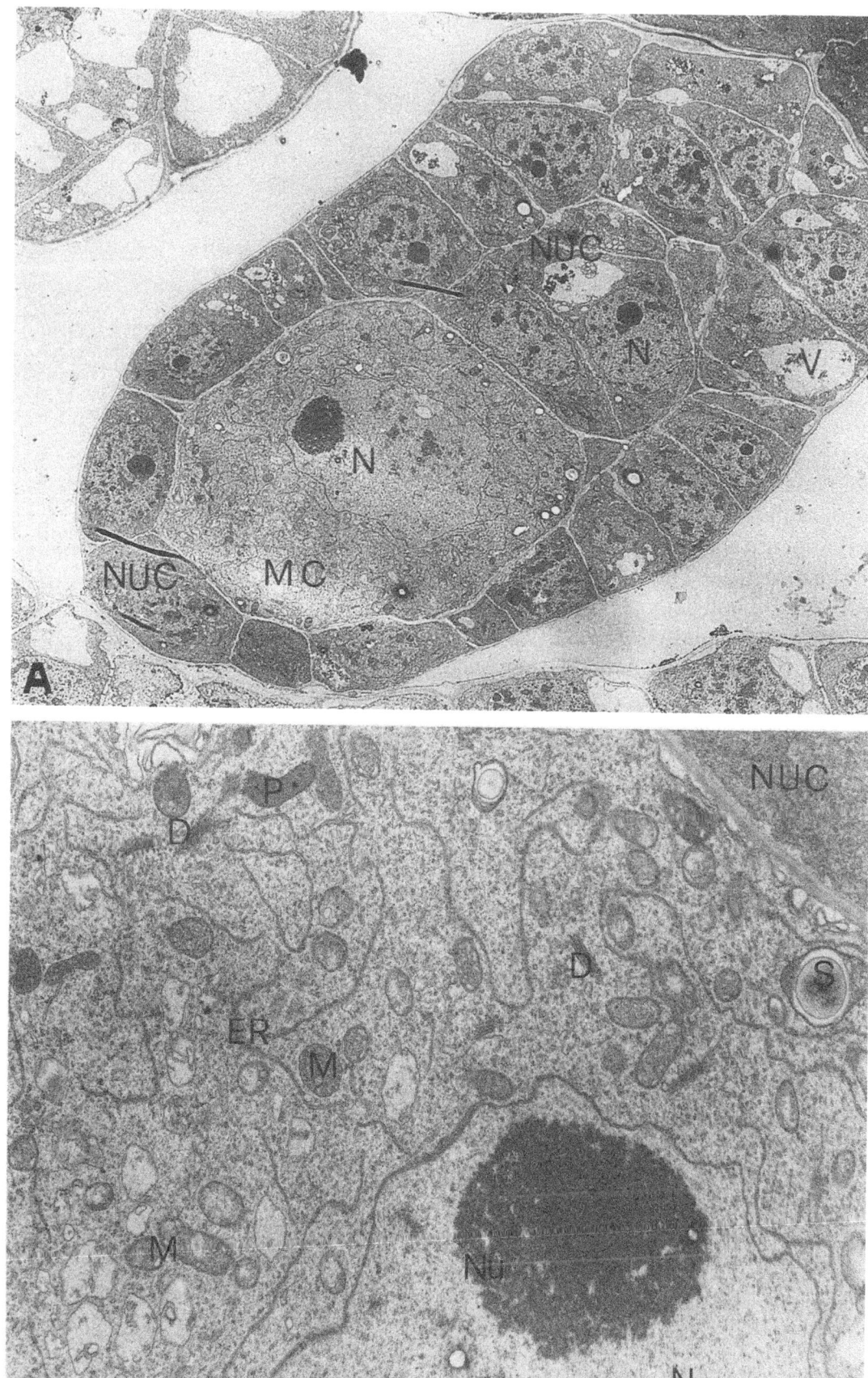

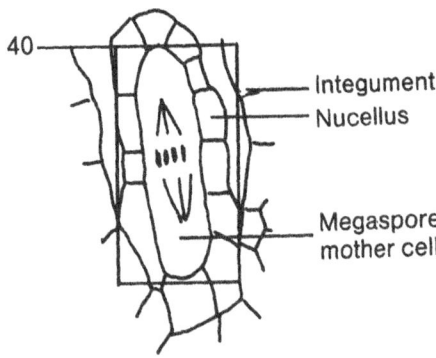

40 — Integument
— Nucellus

Megaspore
mother cell

Plate 40. Megaspore mother cell of *Impatiens walleriana* Hook f. at metaphase I of meiosis. The micropyle is at the top side of the photograph. *Impatiens* has the bisporic type of embryo sac development, and the micropylar dyad cell degenerates shortly after its formation. The longitudinal section of the ovule shows the median plane, and demonstrates the non-equatorial positioning of the metaphase plate (arrows). Note the localization of the endoplasmic reticulum and the uneven distribution of the cell organelles. Most of which are located in the chalazal portion of the cell. The arrowheads point to dictyosomes. x 7,600.

(Plate 40 reproduced by permission from de Boer-de Jeu M (1978) Comm Agr Univ Wageningen 16: 1-128).

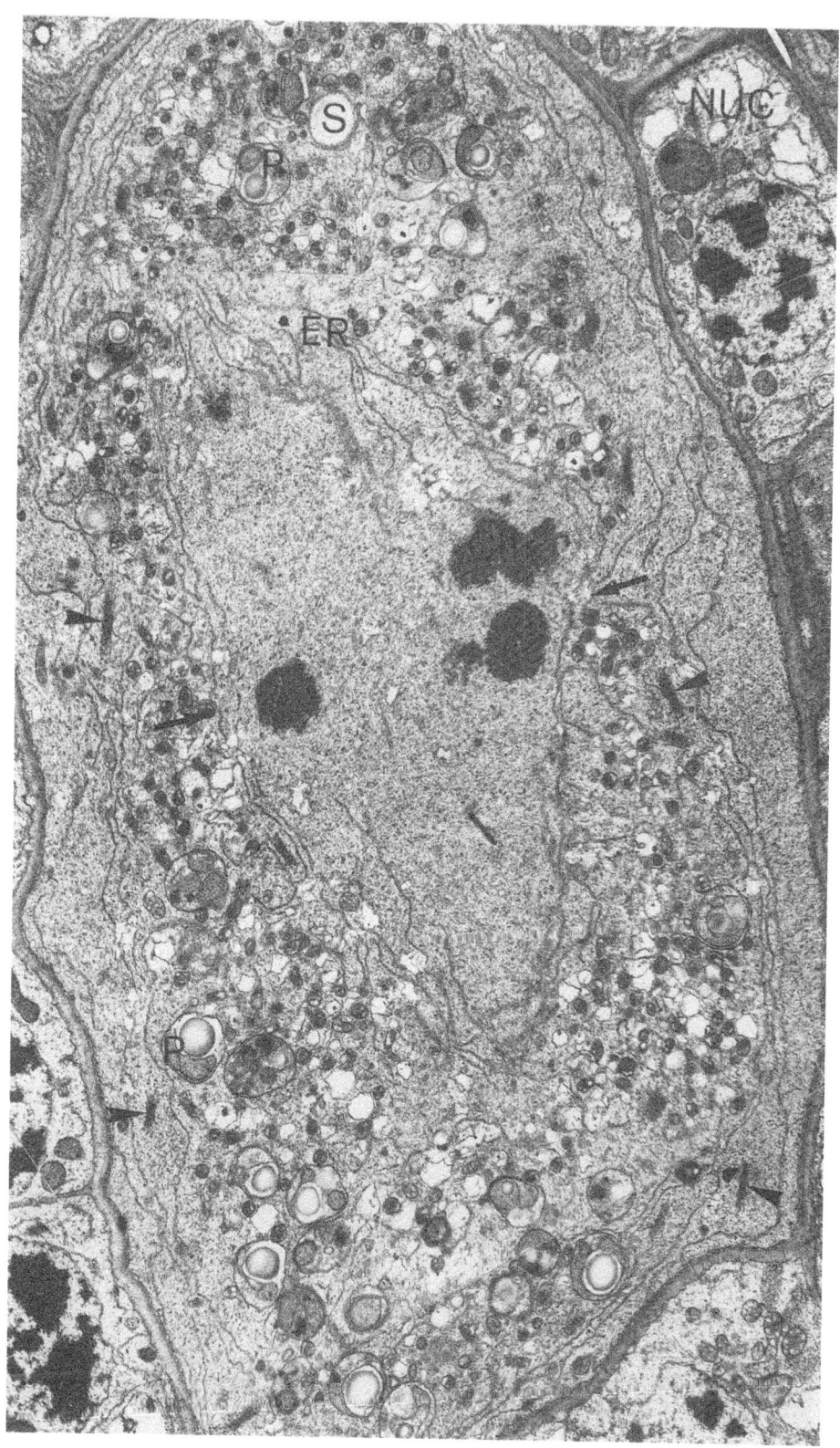

107

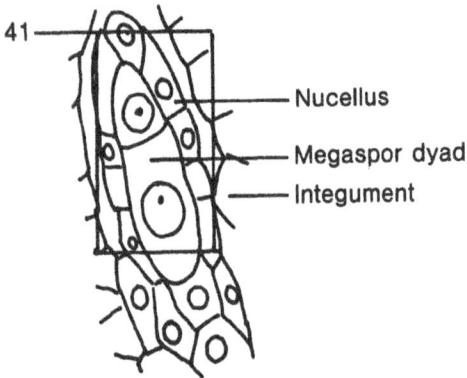

Nucellus

Megaspor dyad

Integument

Plate 41. Cytokinesis after completion of meiosis I in the megaspore mother cell of *Impatiens walleriana* Hook f.. The micropylar pole of the ovule is at the top side of the photograph. As a result of the acentral positioning of the phragmoplast (arrows) cytokinesis results in a small micropylar cell and a large chalazal cell. The uneven distribution of organelles which occurs during metaphase, and the unequal division, determines that most of the cytoplasm and the organelles of the megaspore mother cell are passed to the large, chalazal cell. Arrowheads point to mitochondria. x 7,600.

(Plate 41 courtesy M de Boer-de Jeu, Wageningen).

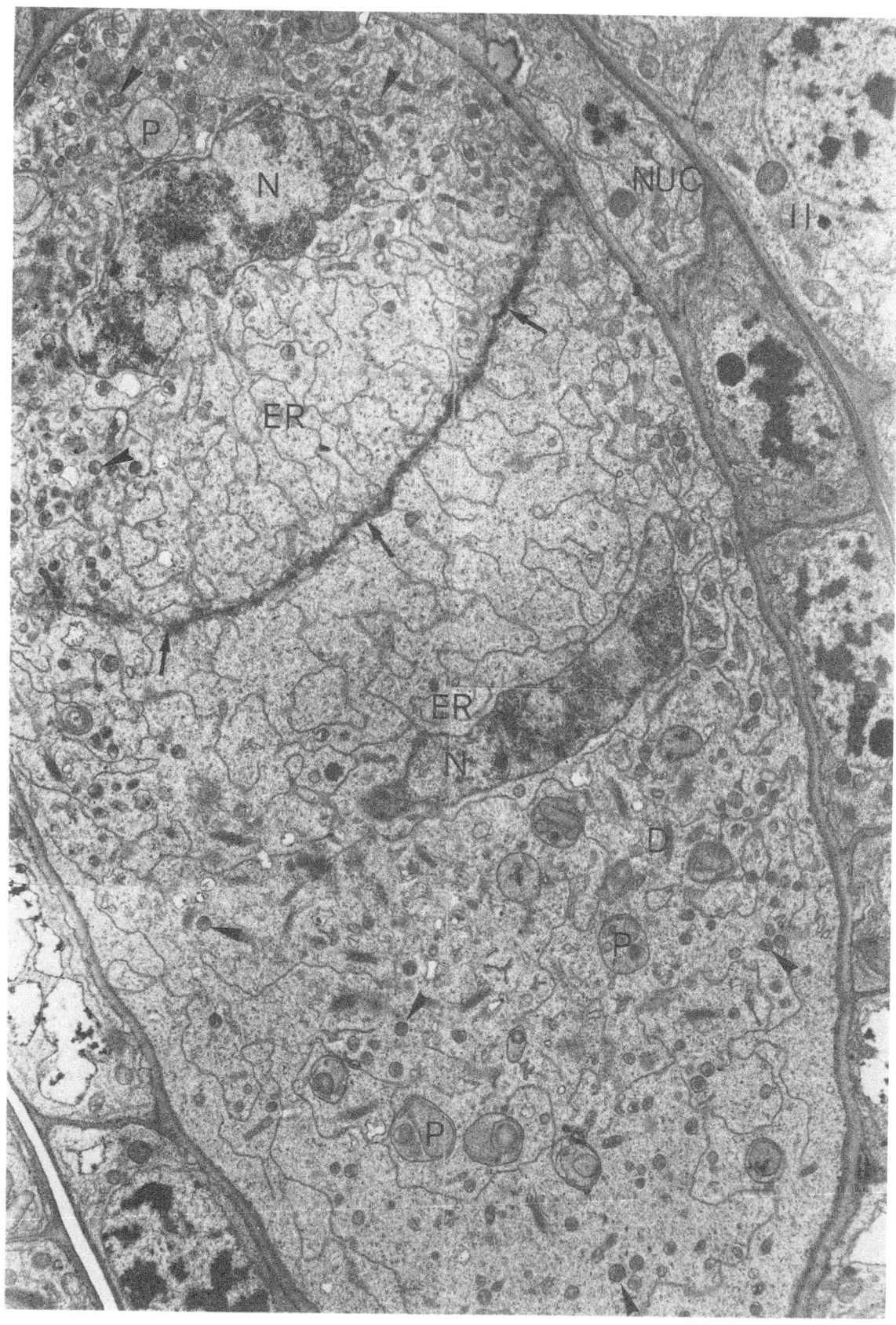

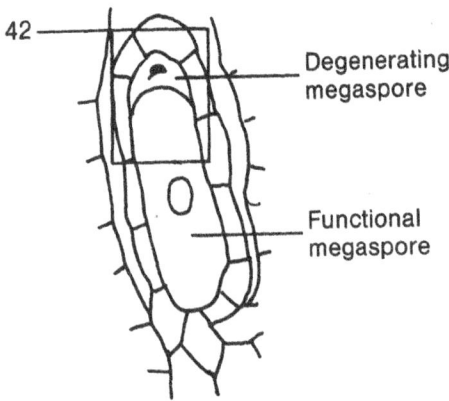

42

Degenerating
megaspore

Functional
megaspore

Plate 42. Cell differentiation after dyad formation in the ovule of *Impatiens walleriana* Hook f.. After its formation the small micropylar cell (at top side of photograph) starts to degenerate. Its cytoplasm and its nucleus become more electron-dense, in comparison to those of the chalazal cell. Whereas the chalazal cell starts to enlarge, the micropylar cell starts to shrink. At this stage of development the organelles of the chalazal dyad cell are again randomly distributed. The nucleus of the chalazal cell will undergo the second meiotic division. x 17,000.

(Plate 42 courtesy M de Boer-de Jeu, Wageningen).

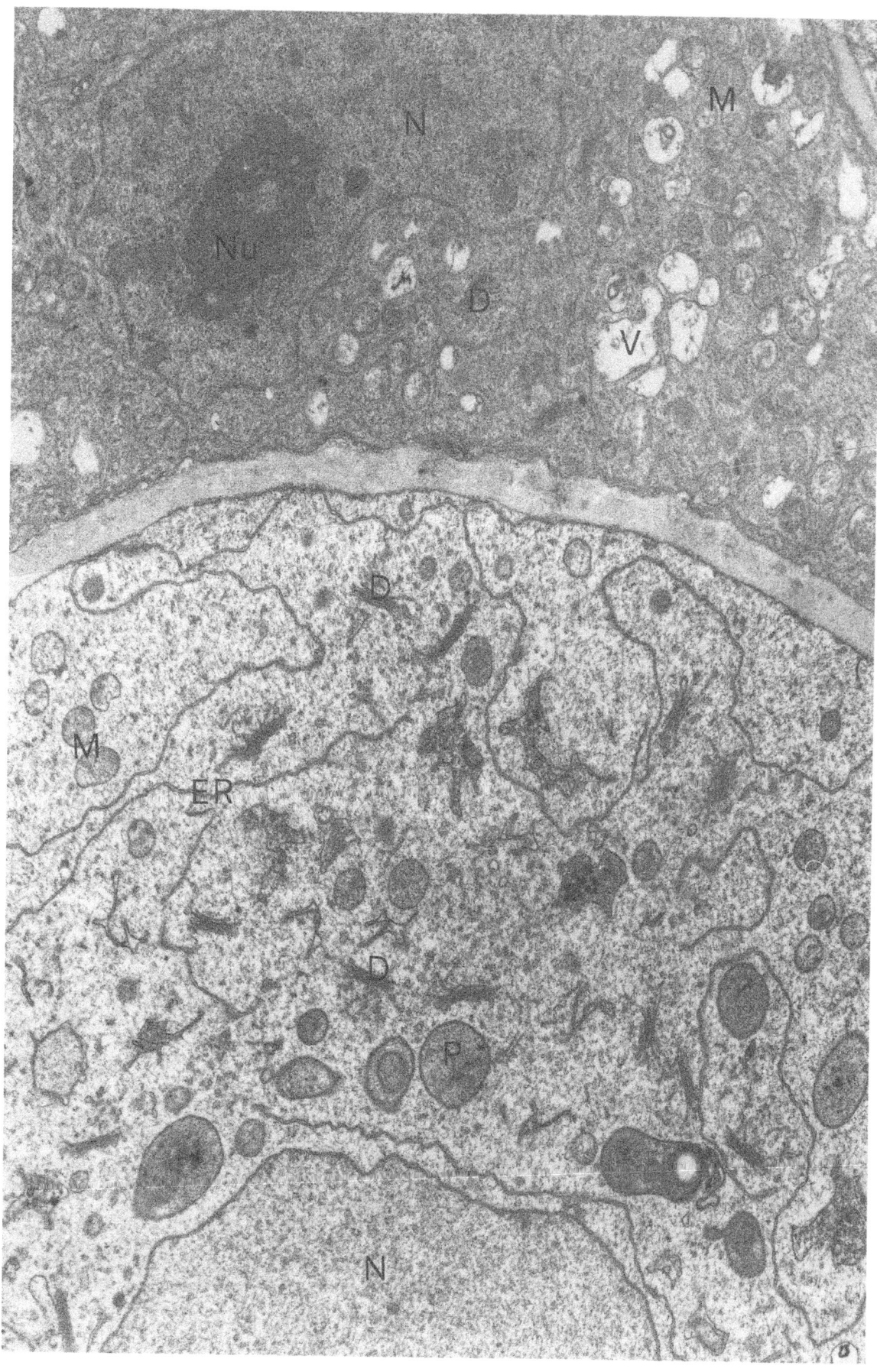

111

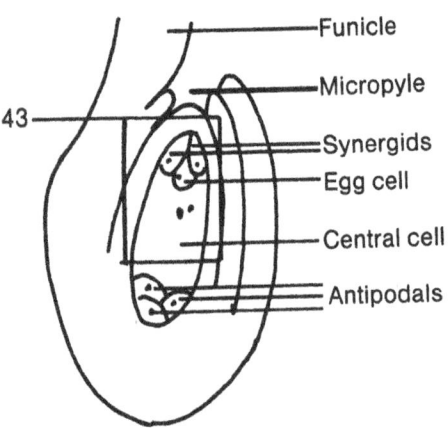

Plate 43. Longitudinal section through the micropylar part of the mature embryo sac of *Spinacia oleracea* L.. The micropyle is positioned at the top of the picture. The micrograph shows one of the synergids, the egg cell and part of the central cell. The micropylar half of the synergid is cytoplasmic rich, whereas the chalazal half is highly vacuolated. The synergid nucleus is located at the base of the cytoplasmic rich half of the cell. Most of the cytoplasm of the egg cell is situated at the chalazal pole of the cell, as is the nucleus.

The synergids and egg cell together form the egg apparatus which is attached to the embryo sac wall only at the extreme micropylar position. The rest of the egg apparatus is protuding into the central cell. The central cell contains the two polar nuclei, which are positioned close to each other near the chalazal pole of the egg apparatus. x 2,500.

(Plate 43 reproduced by permission from Wilms H (1981) Acta Bot Neerl 30: 75-99).

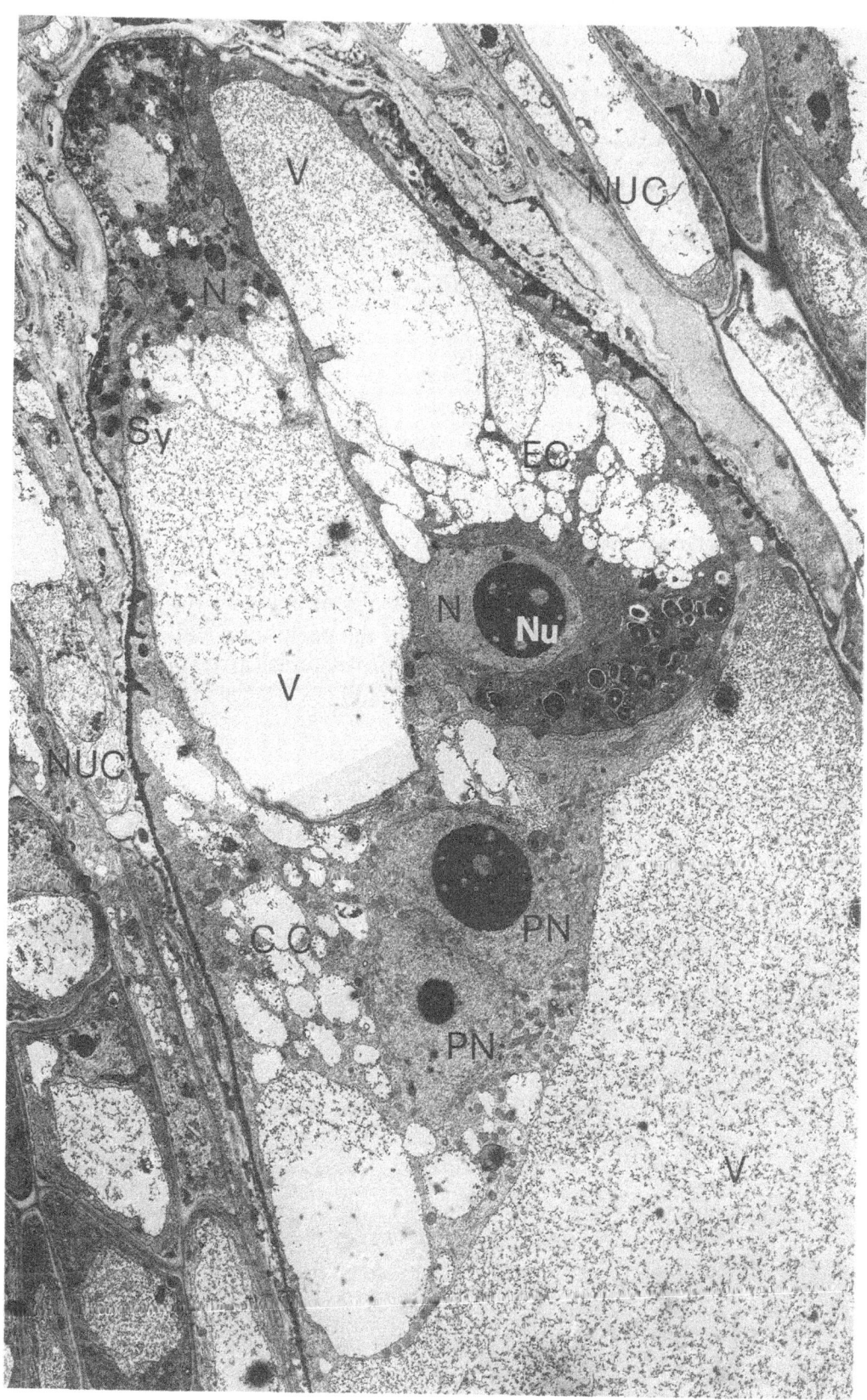

113

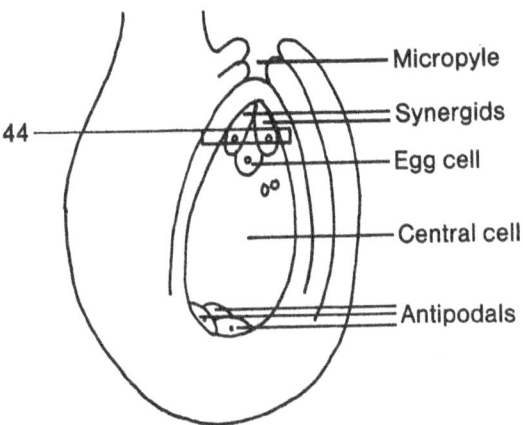

Plate 44. Transverse section through the micropylar portion of the embryo sac of *Spinacia oleracea* L., showing the two synergids and the egg cell in triangular arrangement. The walls in between the cells in this region are very thin. The portion of the peripheral wall of the central cell near the egg apparatus has protrusions (arrows) and can be characterized as a transfer wall. x 2,400.

(Plate 44 courtesy H Wilms, Wageningen).

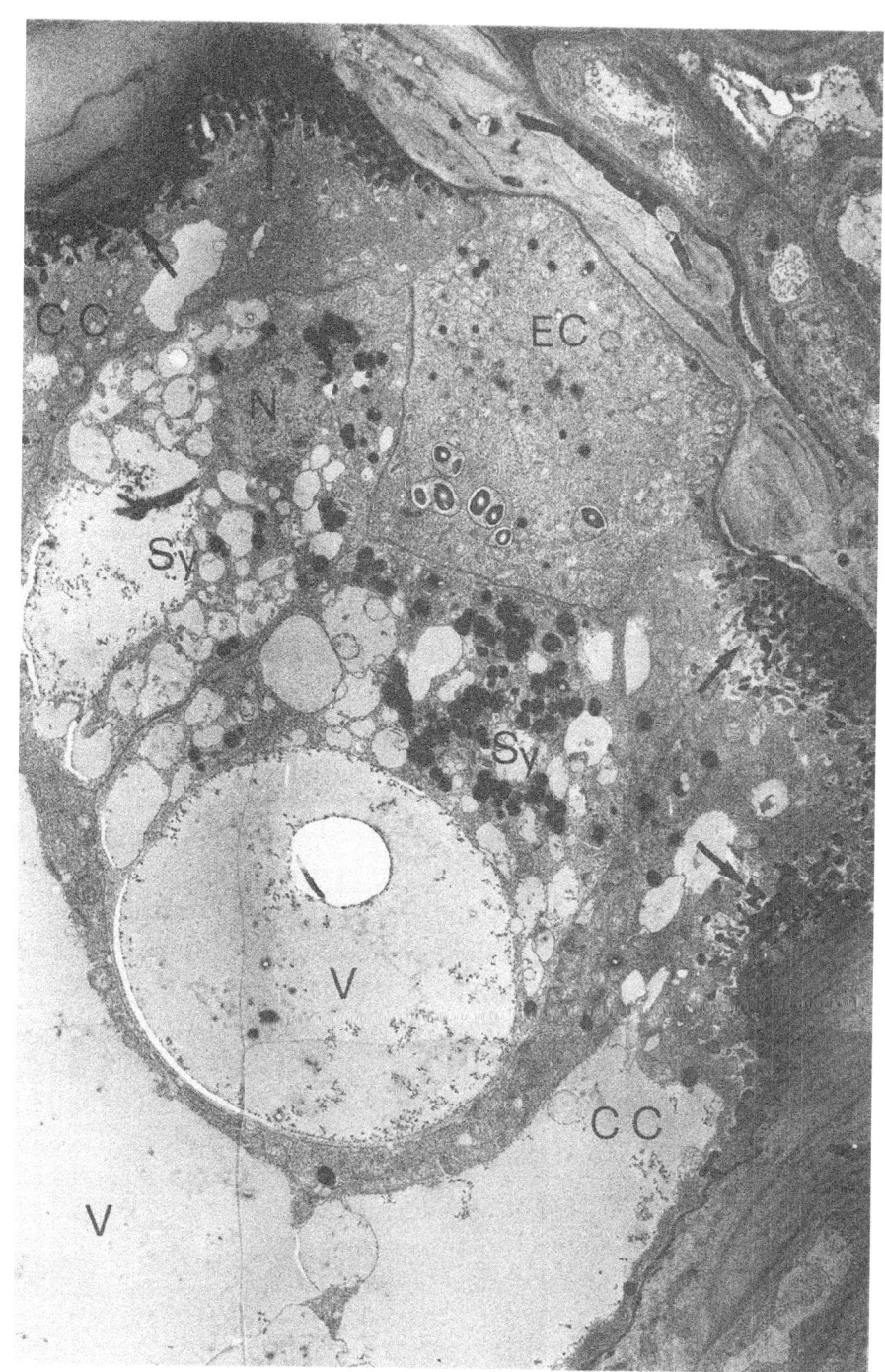

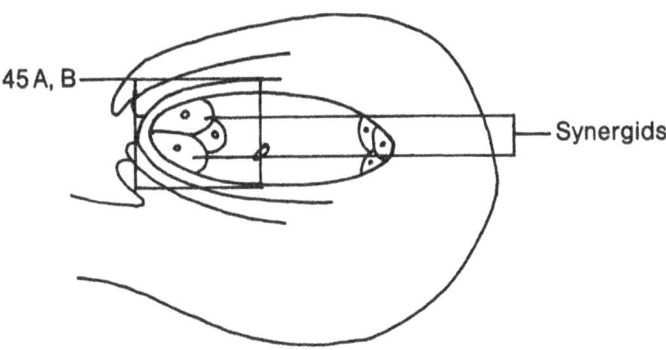

Plate 45. Longitudinal sections through the micropylar region of the embryo sac of *Cucumis sativus* L..

A. shows the synergids and part of the central cell just after cell formation. The ovule of *Cucumis sativus* is crassinucellate. The space for the developing embryo sac results from the degeneration of neighbouring nucellus cells (arrows). At this young stage of development the synergids are completely surrounded by cell walls. x 1,800.

B. shows the same part of the embryo sac at the mature, receptive stage. The synergids have strongly enlarged. At the micropylar pole a filiform apparatus has been formed, consisting of an irregularly thickened, labyrinthous cell wall. At the chalazal portion of the synergids only a plasma membrane is present. The micropylar two-third portion of the synergids is completely filled with cytoplasm, while the chalazal one-third is highly vacuolated. Stacks of endoplasmic reticulum are prominent near the nuclei. x 1,800.

(Plates 45A,B courtesy C Theunis, Wageningen).

CC
Sy
N
Sy
N
V
NUC
A

FA
Sy
Sy
N
ER
NUC
N
V
CC
N
V
ER
B

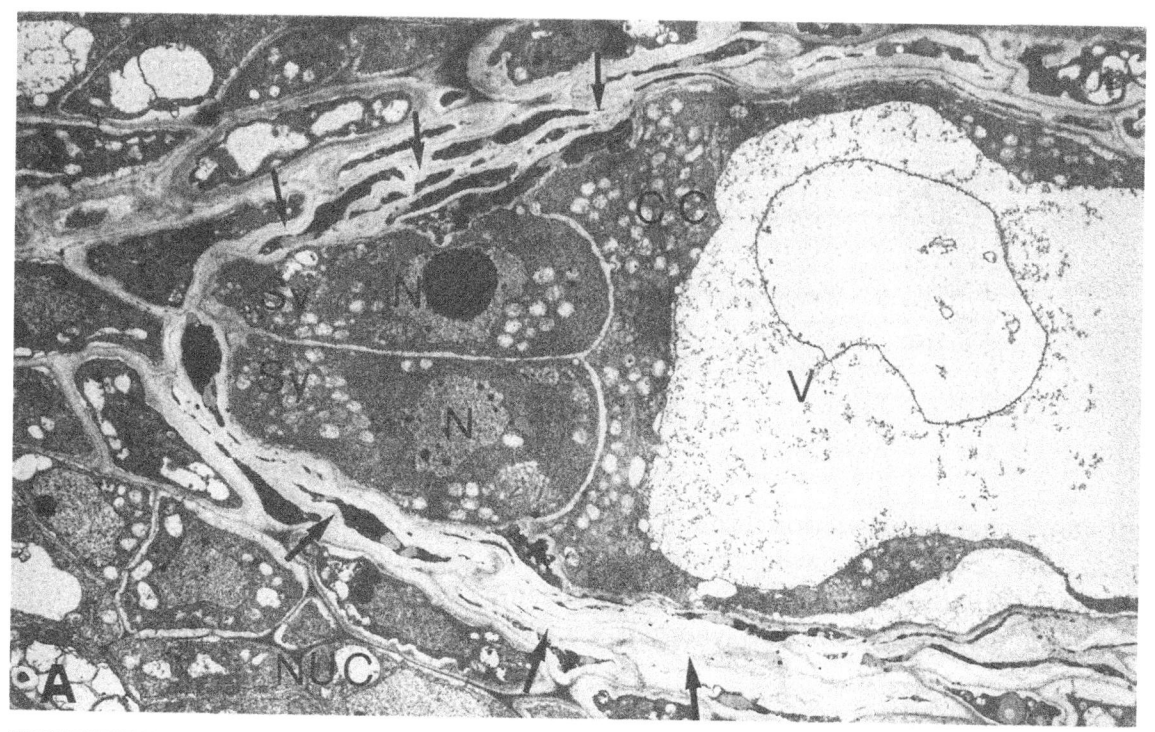

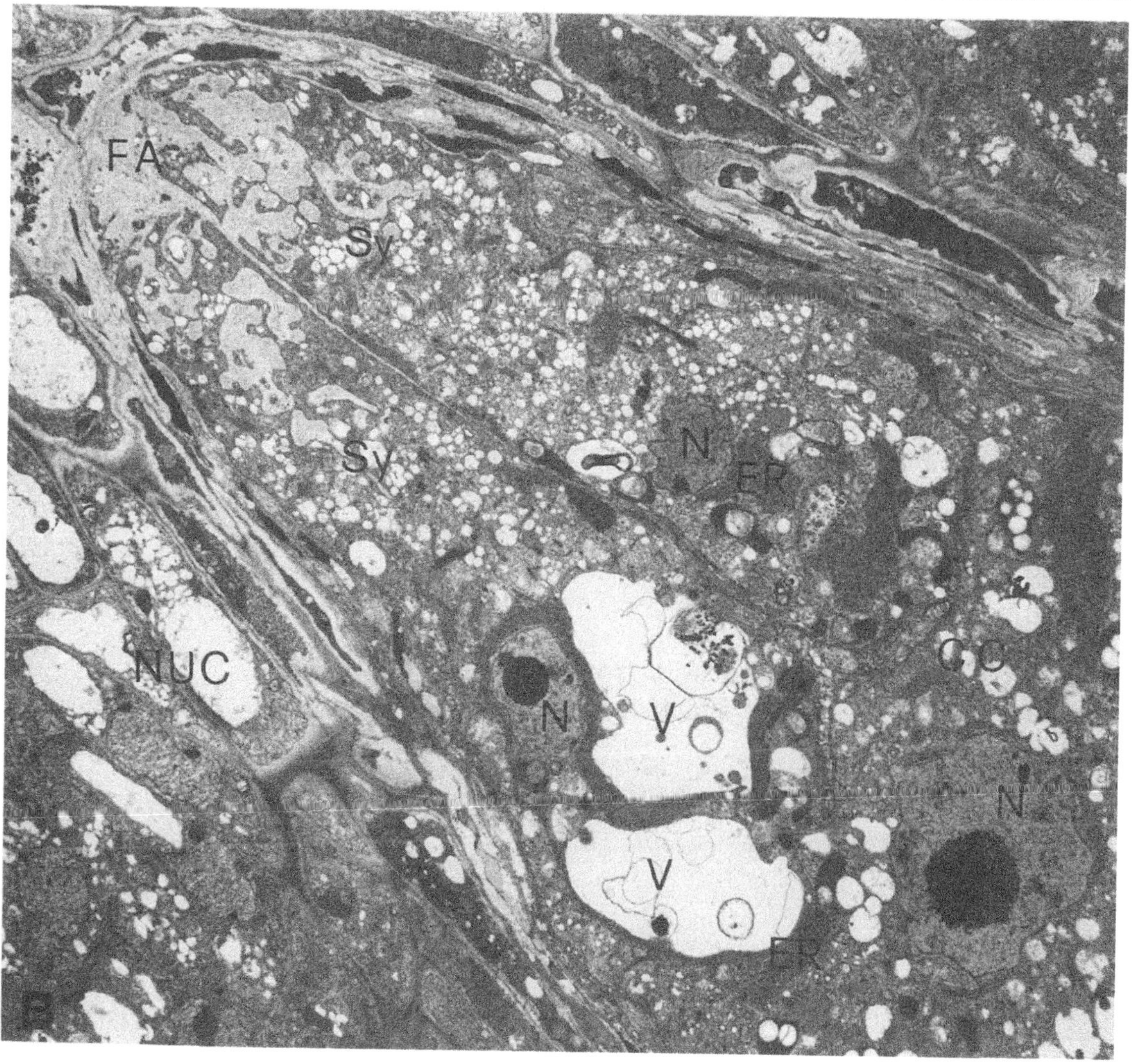

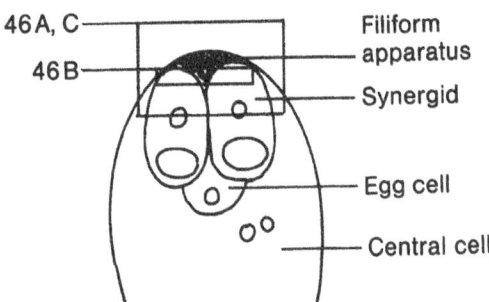

Plate 46A. Longitudinal section through the embryo sac of *Petunia hybrida* Hort. ex Vilm., showing the wedge-shape of the filiform apparatus. The filiform apparatus is a local thickening of the wall at the top and in between the two synergids. x 7,500.

Plate 46B. Transverse section through the filiform apparatus of *Petunia hybrida* Hort. ex Vilm., which shows the loosely organized cellulosic network. Before embedding in butylmetacrylate the wall matrix of the filiform apparatus was removed by short treatment with a mixture of acetic acid and hydrogen peroxide. This treatment did not remove the matrix in the other cell walls (asterisk). After sectioning the butyl methacrylate was removed, and the remaining material was shadowed with platinum. x 12,000.

Plate 46C. Longitudinal section through the embryo sac of *Spinacia oleracea* L., revealing the transfer wall-like structure of the filiform apparatus. In *Spinacia* the filiform apparatus consists of local, irregular thickenings of the micropylar portions of the peripheral synergid walls. The wall in between the synergids (arrows) is not thickened. x 24,000.

(Plates 46A, B reproduced by permission from Van Went J (1970a) Acta Bot Neerl 19: 121-132; Plate 46C courtesy H Wilms, Wageningen).

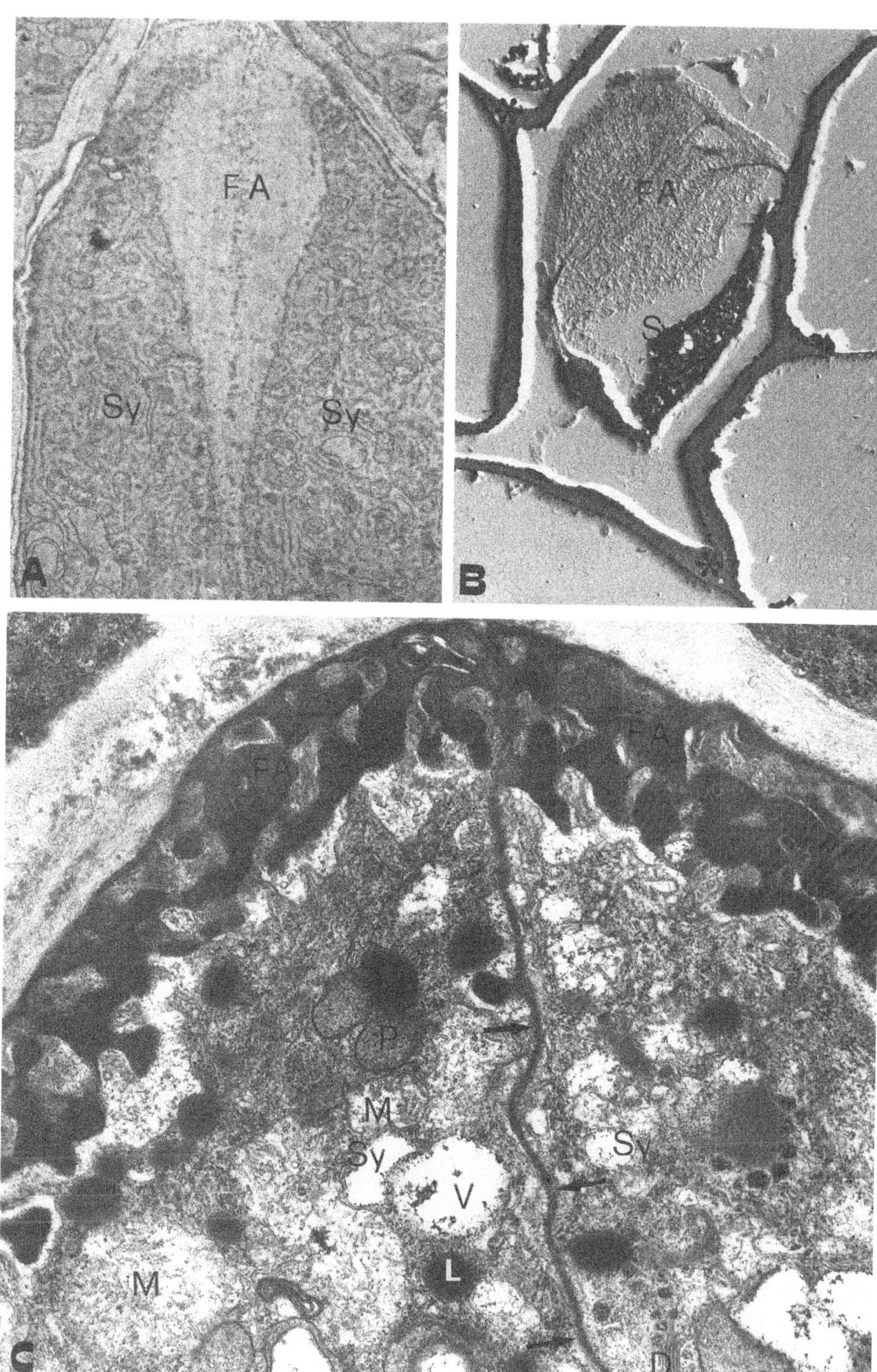

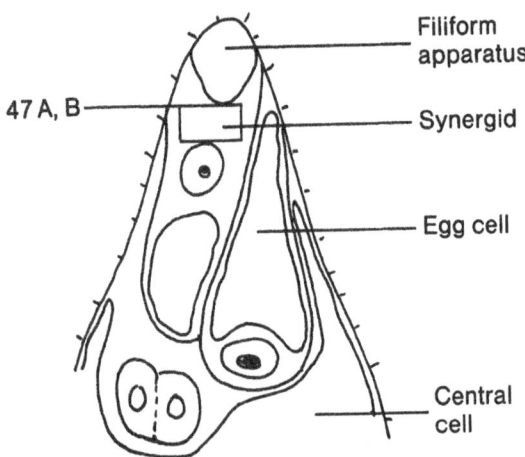

Plate 47A. Ultrastructure of the synergid cytoplasm of *Brassica campestris* L. at the mature embryo sac stage. Most prominent are the extensive rough endoplasmic reticulum, and the dictyosomes producing numerous large Golgi-vesicles. x 33,000.

Plate 47B. In *Brassica campestris* L. both synergids start to degenerate shortly before the arrival of the pollen tube. Degeneration is marked by the deposition of electron-dense materials onto the membranes of the mitochondria and the endoplasmic reticulum. Apparently, degeneration in one synergid is more advanced then in the other. x 19,000.

(Plates 47A, B reproduced by permission from J Van Went, M Cresti (1988a) Sex Plant Reprod 1: 208-216).

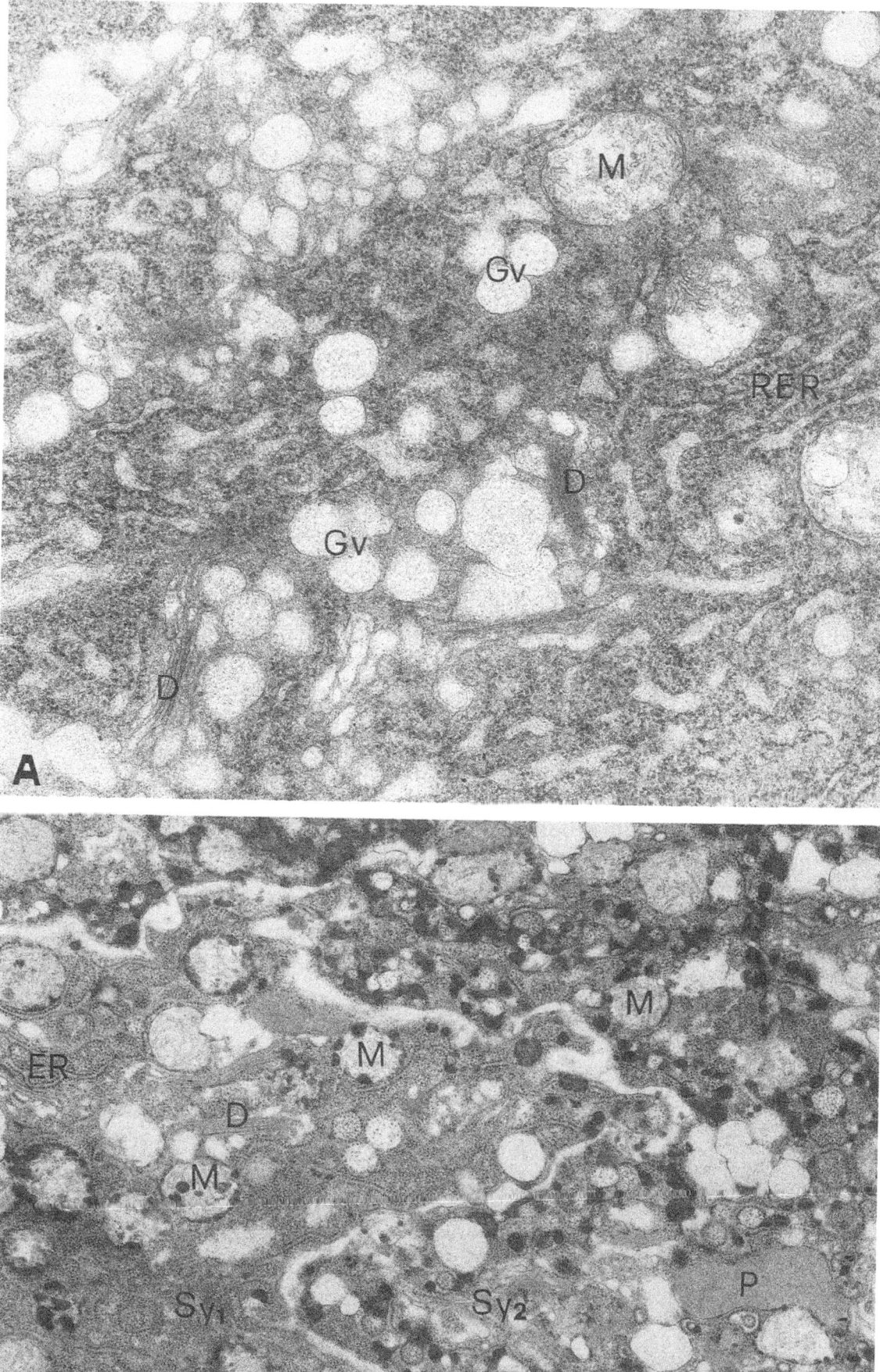

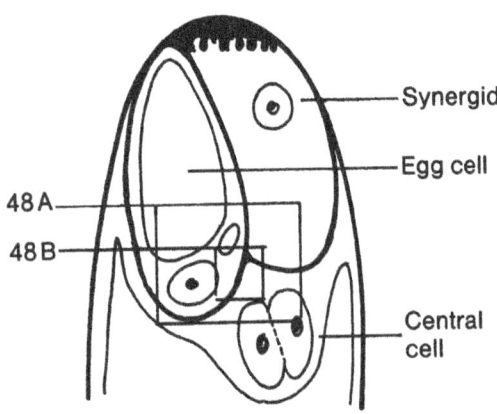

Plate 48A. Longitudinal section through the mature embryo sac of *Brassica oleracea* L., shortly before arrival of the pollen tube. (Same specimen as shown in plate 46). The synergids have already started to degenerate, and some synergid cytoplasm has intruded in between the egg cell and the central cell. The regular boundary between the egg cell and the central cell is very thin (indicated by arrows). x 9,000.

Plate 48B. Enlarged portion of plate 48A. The double arrows indicate the presence of two membranes in between the synergid and the neighbouring cells. The synergid content in between the egg cell and the central cell is bordered by single membranes, showing that the synergid membrane has been disrupted, resulting in the release of the cytoplasmic contents. x 21,000.

(Plates 48A, B reproduced by permission from Van Went J, Cresti M (1988a) Sex Plant Reprod 1: 208-216).

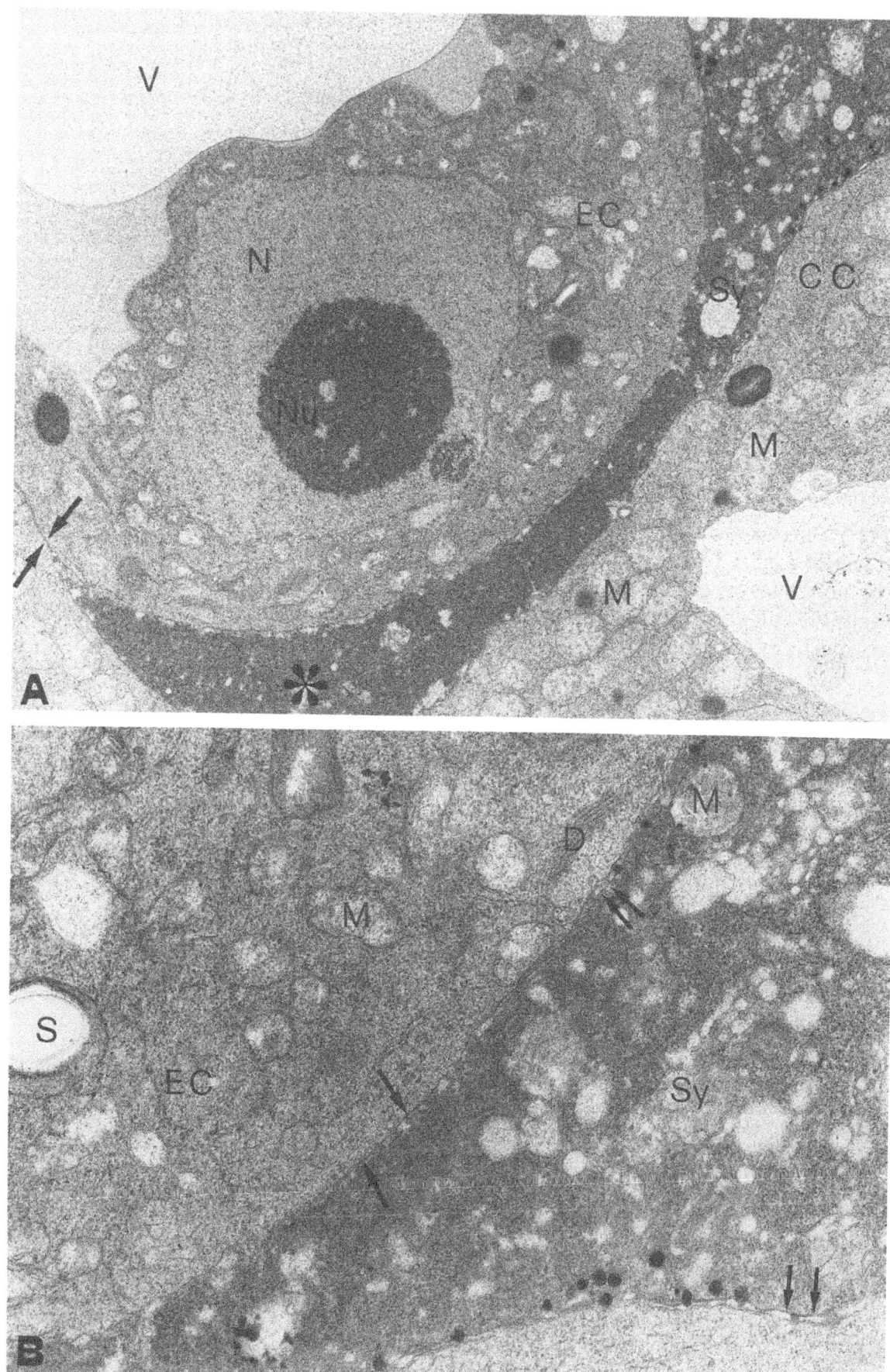

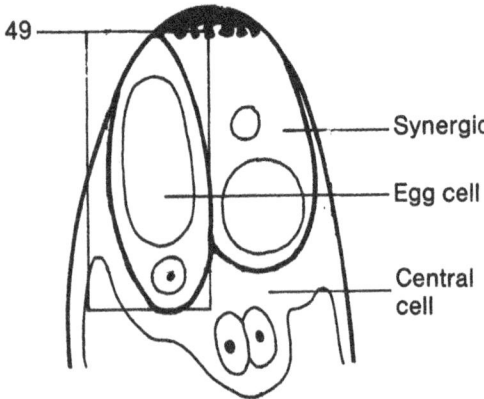

Plate 49. Longitudinal section through the mature embryo sac of *Spinacia oleracea* L., showing the egg cell. The micropylar part of the cell is highly vacuolated. Usually, the egg nucleus is located in the chalazal part of the cell. This portion of the egg cell is surrounded by a very thin cell wall, or by the plasma membrane only. Most of the cell organelles are positioned, like the nucleus, in the chalazal part of the egg cell. x 8,700.

(Plate 49 courtesy H Wilms, Wageningen).

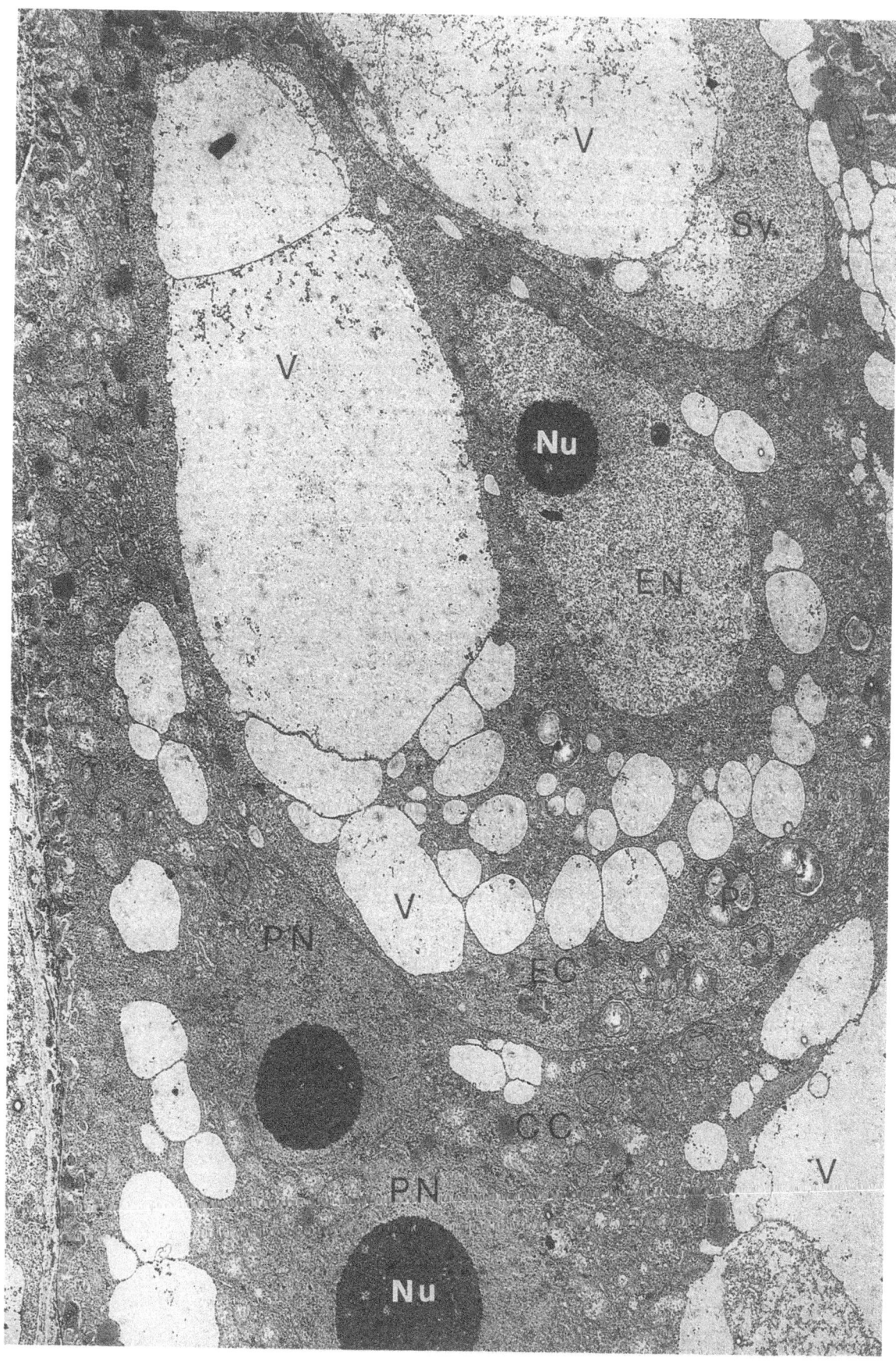

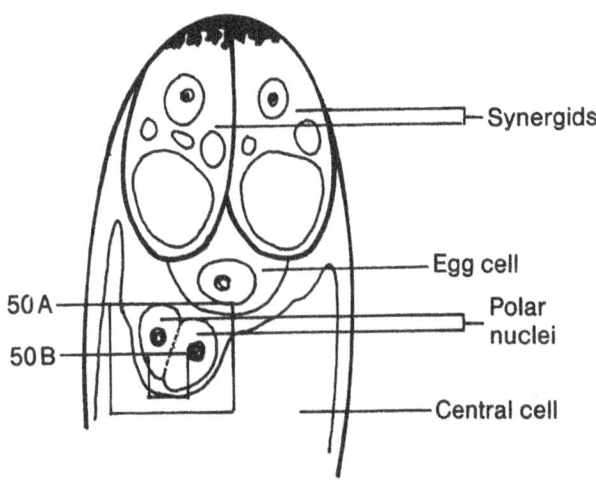

Plate 50A. Longitudinal section through the embryo sac of *Spinacia oleracea* L., showing the two polar nuclei in the central cell. In *Spinacia*, when the embryo sac reaches maturity, the two polar nuclei become positioned close to each other and form protrusions (arrows) as a prelude to the process of nuclear fusion. x 9,100.

Plate 50B. Enlarged portion of the two polar nuclei with protrusions (arrows) of *Spinacia oleracea* through which they contact each other and eventually fuse. At the contact sites the outer nuclear membranes fuse first, followed by fusion of the inner membranes. In this way several bridges between the nuclei are formed that subsequently enlarge and coalesce. x 38,000.

(Plate 50A courtesy H Wilms, Wageningen; Plate 50B reproduced by permission from H Wilms (1981a) Acta Bot Neerl 30: 75-99).

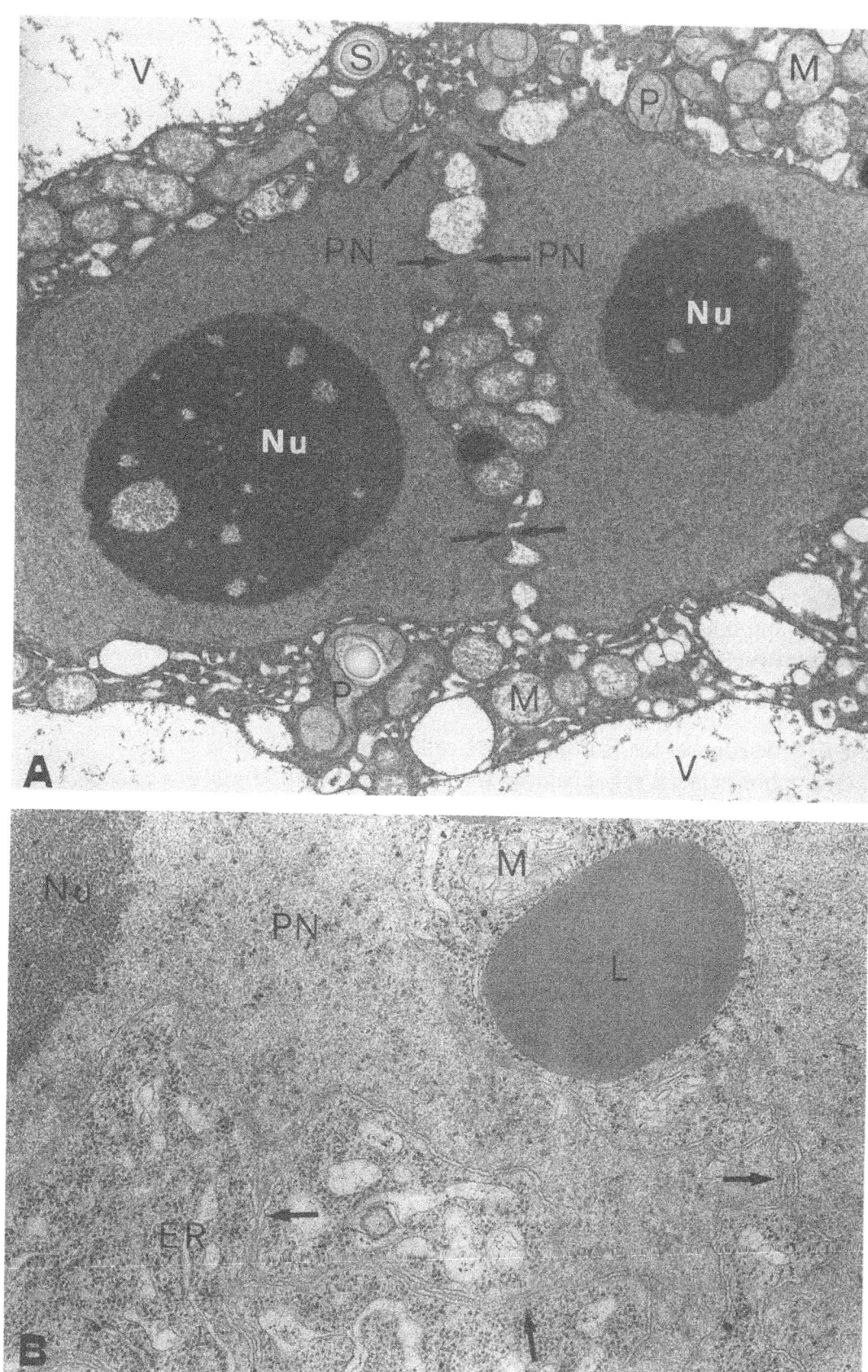

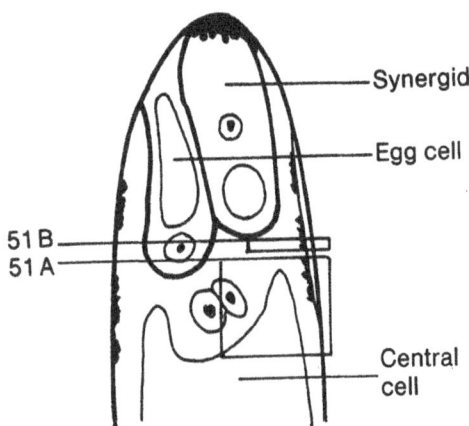

Plate 51A. Portion of the central cell of *Spinacia oleracea* L. from a mature embryo sac showing the ultrastructure of the cytoplasm near the polar nuclei. There is an extensive system of rough endoplasmic reticulum and many mitochondria. The plastids show a relatively simple substructure and contain small starch grains. x 10,000.

Plate 51B. Portion of the mature central cell of *Spinacia oleracea* L. in the region near the egg apparatus. In this region the peripheral cell wall of the central cell has a very irregular inner surface, due to the presence of many cell wall extensions (arrow). Furthermore, the wall is strongly stained, in comparison to the walls of the surrounding nucellus cells. x 3,800.

(Plates 51A,B courtesy H Wilms, Wageningen).

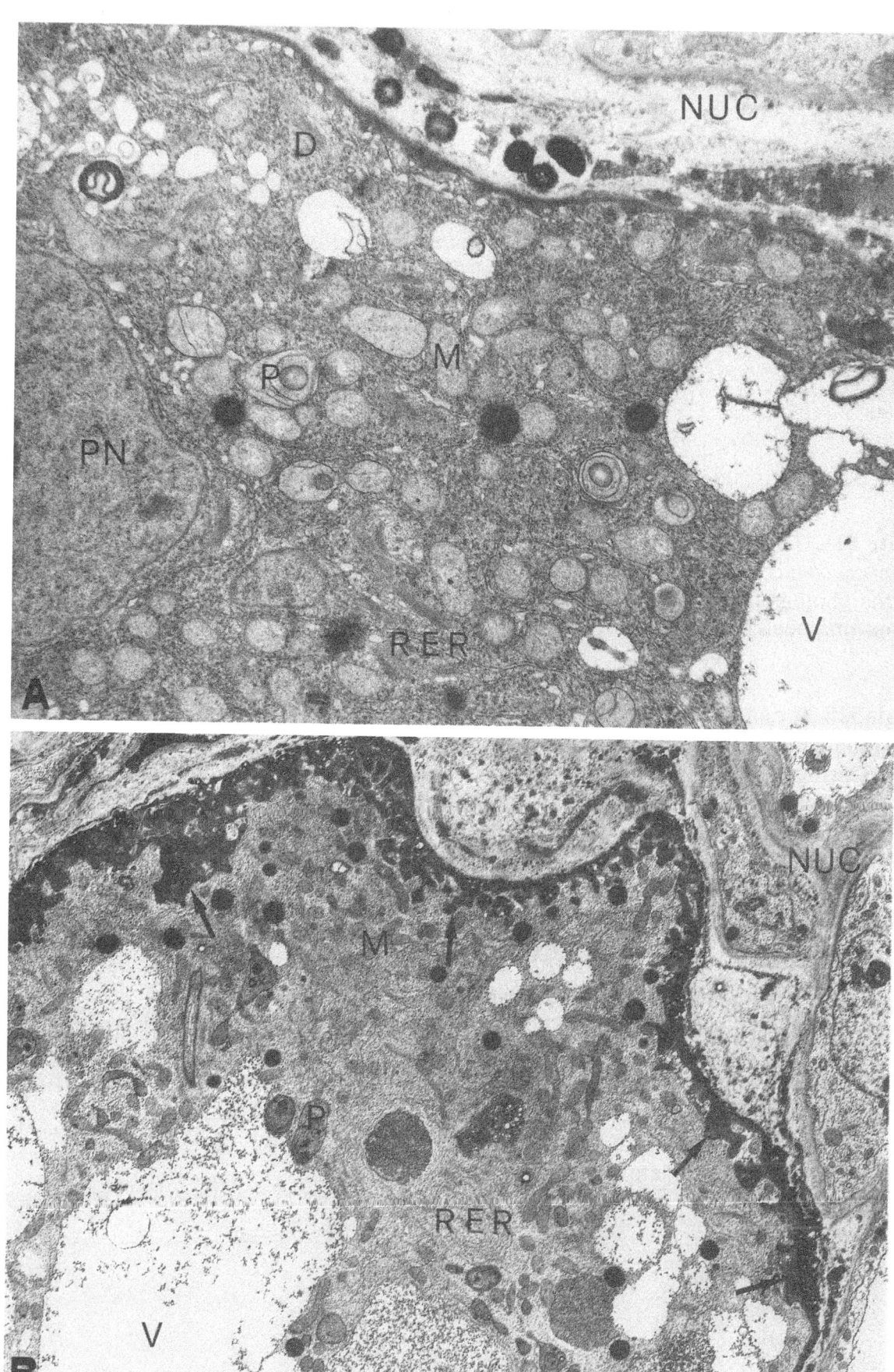

129

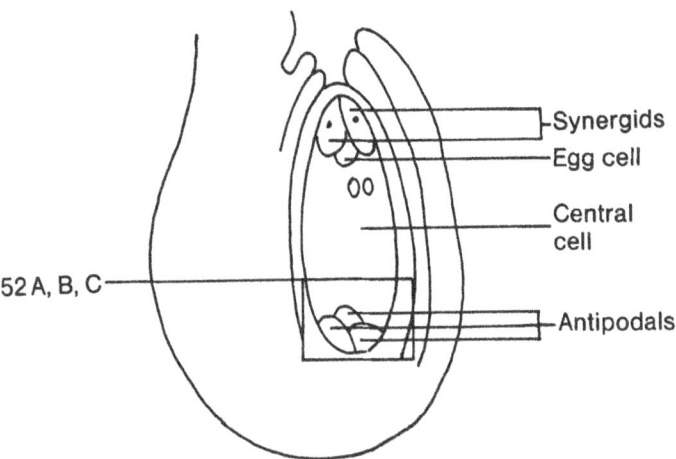

Plate 52A. The antipodal cells of the developing embryo sac of *Spinacia oleracea* L. The neighbouring central cell is at the left side of the photograph. During the maturation of the embryo sac the three antipodals degenerate. x 2,800.

Plate 52B. The antipodal cells of the mature embryo sac of *Zea mays* L. Initially three antipodal cells are formed in the embryo sac of *Zea*. During the embryo sac maturation these divide and eventually an antipodal tissue, composed of approximately twenty cells is formed. Some of the cells have several nuclei, because of absence of cytokinesis or the only partial formation of cell walls (arrow). The neighbouring central cell is at the lower right corner of the photograph. x 1,000.

Plate 52C. Enlarged portion of an antipodal cell of *Zea mays* L. adjacent to the central cell. x 8,000.

(Plate 52A courtesy H Wilms, Wageningen; Plates 52B reproduced by permission from A Van Lammeren (1986b) Agric Univ Wageningen Pap 86-1; Plate 52C original J Van Went).

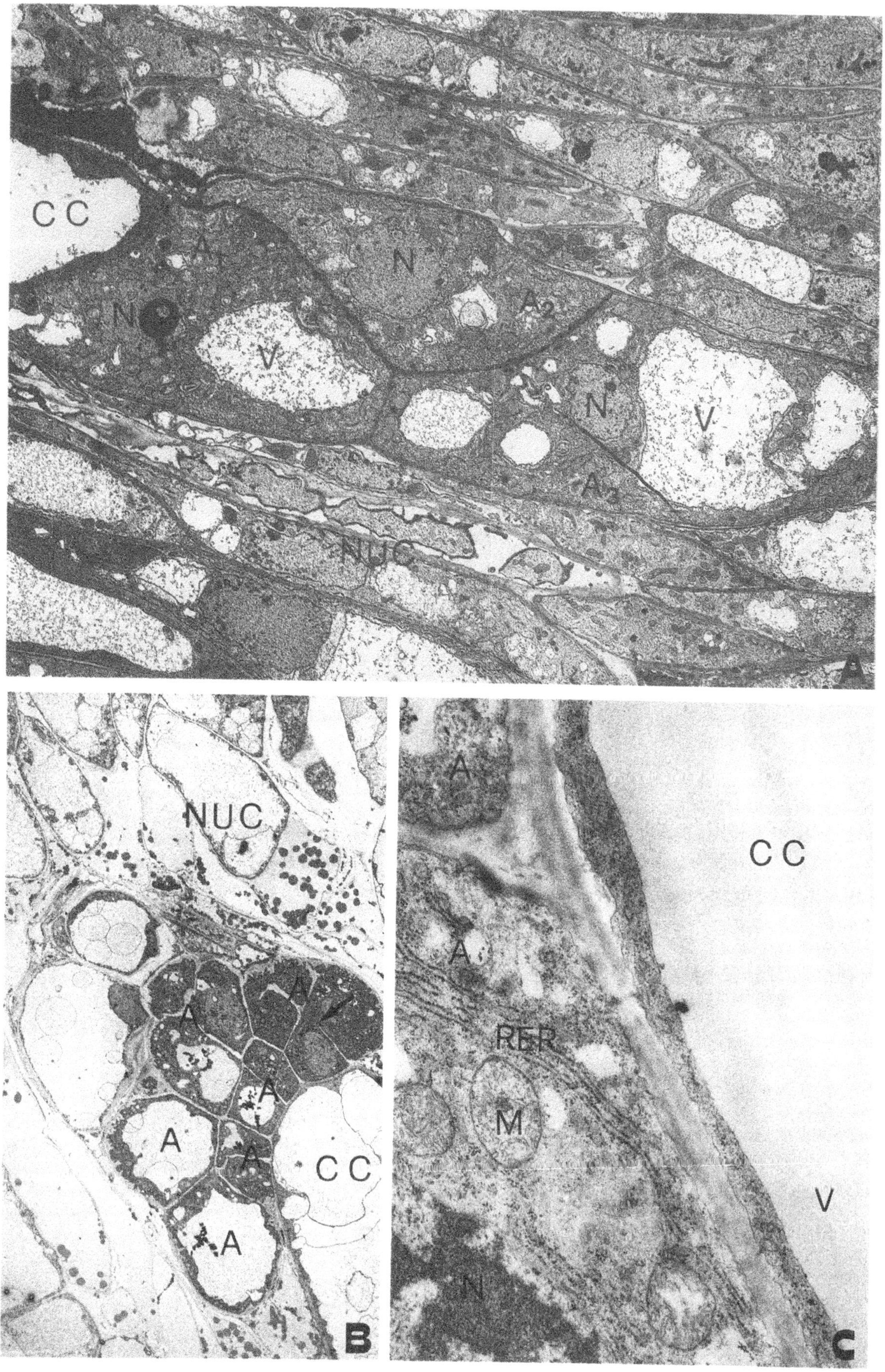
CC
A_1
N
N
V
A_2
N
V
A_3
NUC
A
NUC
A
A
A
A
A
CC
A
B
A
CC
A
RER
M
V
N
C

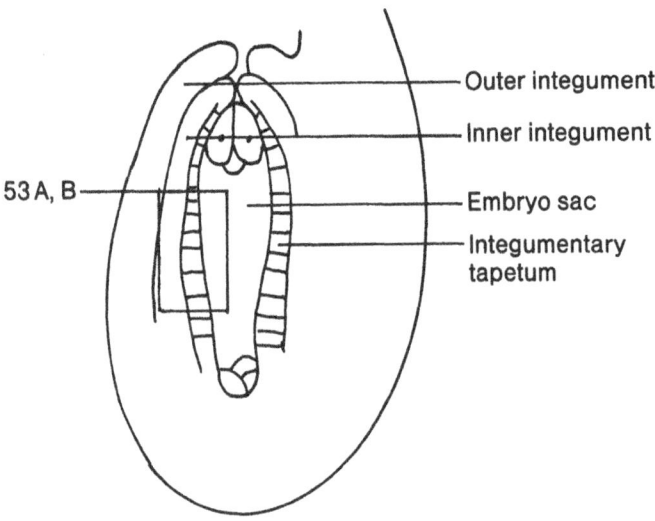

Outer integument

Inner integument

53 A, B

Embryo sac

Integumentary
tapetum

Plate 53A. Longitudinal section through the mature ovule of
Impatiens walleriana Hook f. The cells of the integument epidermis
bordering the embryo sac, develop into a characteristic layer, called
the integumentary tapetum or endothelial layer. The elongated
cells are oriented perpendicular to the long axis of the embryo sac.
During development, electron-dense and acetolysis-resistant
material is deposited in the endothelium wall bordering the embryo
sac (arrows). x 6,400.

Plate 53B. Longitudinal section through the mature ovule of
Impatiens glandulifera Royle, showing the very pronounced
endothelial layer bordering the embryo sac. x 3,500.

(Original J Van Went).

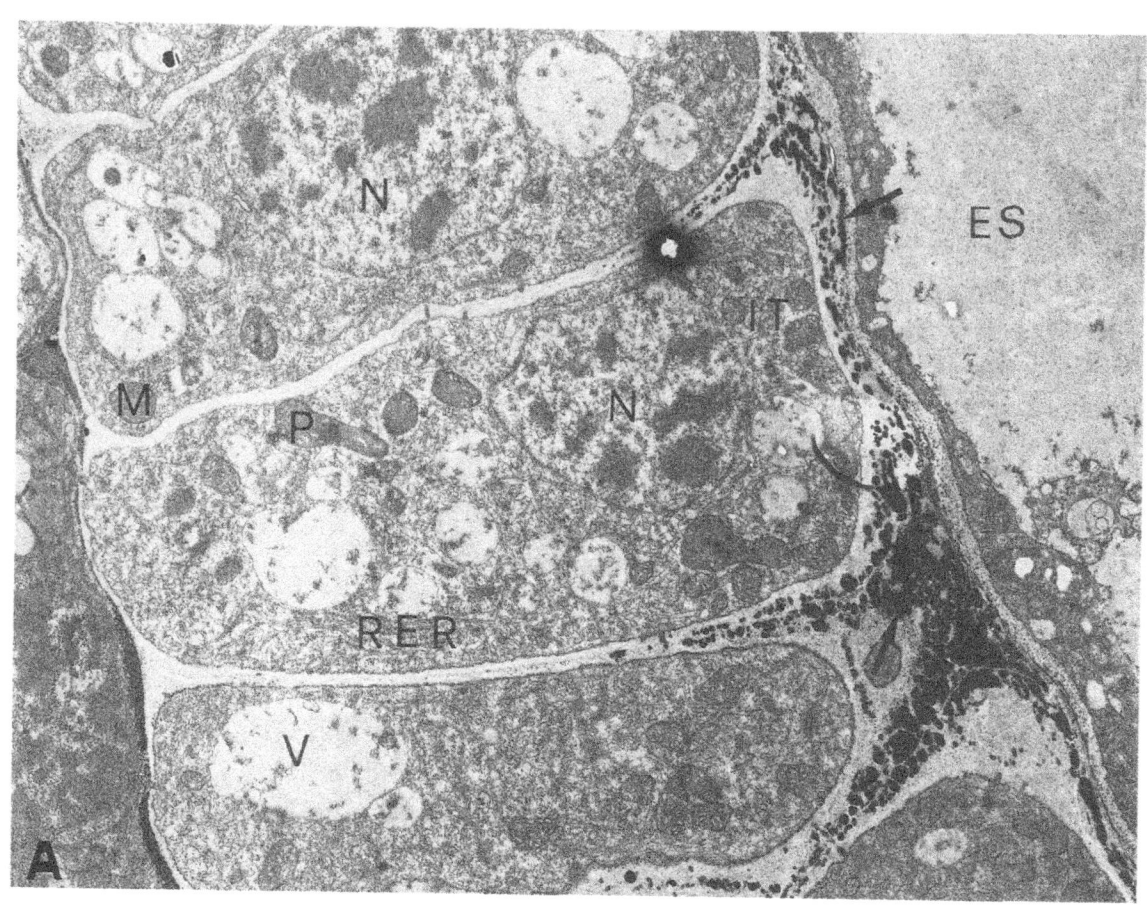

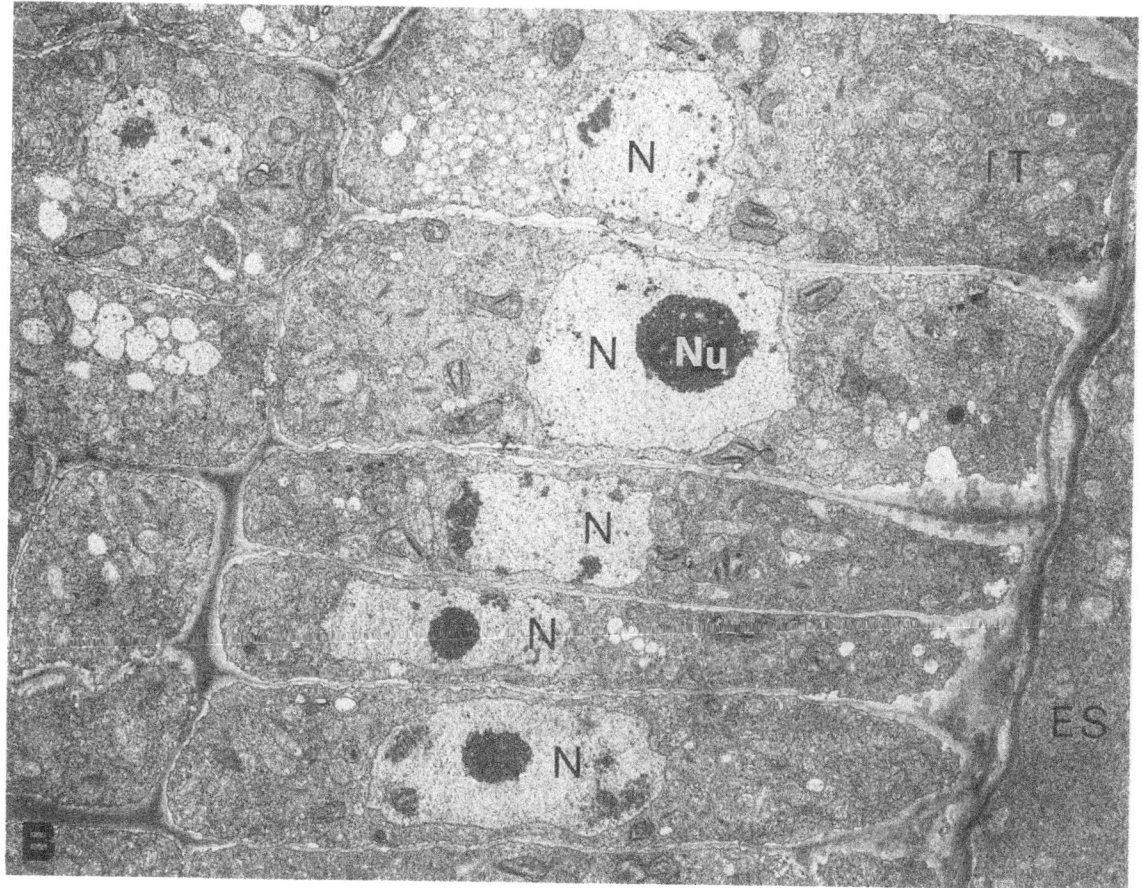

Plate 54. Enzymatic isolation of living embryo sacs of *Petunia hybrida* Hort. ex Vilm. To isolate the embryo sacs dissected ovules are incubated for 2 hrs in a maceration medium containing 3% driselase, 0.1% MES buffer, pH 5.5, and 8% mannitol. Plates A-I show the subsequent stages of the maceration process.

A. After 1 hr of incubation the ovule has become transparent, and the living embryo sac can be observed directly with differential interference contrast microscopy. The egg cell and central cell are in focus. x 600.

B/C. After 2 hrs of incubation a considerable part of the sporophytic tissue is already removed. B. and C. are views of the same ovule but in different focal planes, showing the synergids, the egg cell and central cell. x 900.

D/E. Gentle agitation of the maceration mixture after 2 hrs of incubation results in the rapid and complete separation of sporophytic and gametophytic tissue. Most resistant to separation is the sporophytic tissue called hypostase, which is located at the chalazal side of the embryo sac. During the 2 hrs incubation the cells of the embryo sac retain their original shape, position, and organization. x 900.

F/G. Prolongation of the incubation after 2 hrs results in the gradual transformation of the gametophytic cells into protoplasts. The synergids, egg cell, central cell and antipodal cells become spherical in shape. x 900.

H/I. Ultimately the embryo sac breaks up in individual, spherical protoplasts: synergids, egg cell, central cell and antipodal cells as seen in different focal planes. x 900.

(Plates 54A-I reproduced by permission from Van Went J, Kwee H S (1990) Sex Plant Reprod 3: 257-262).

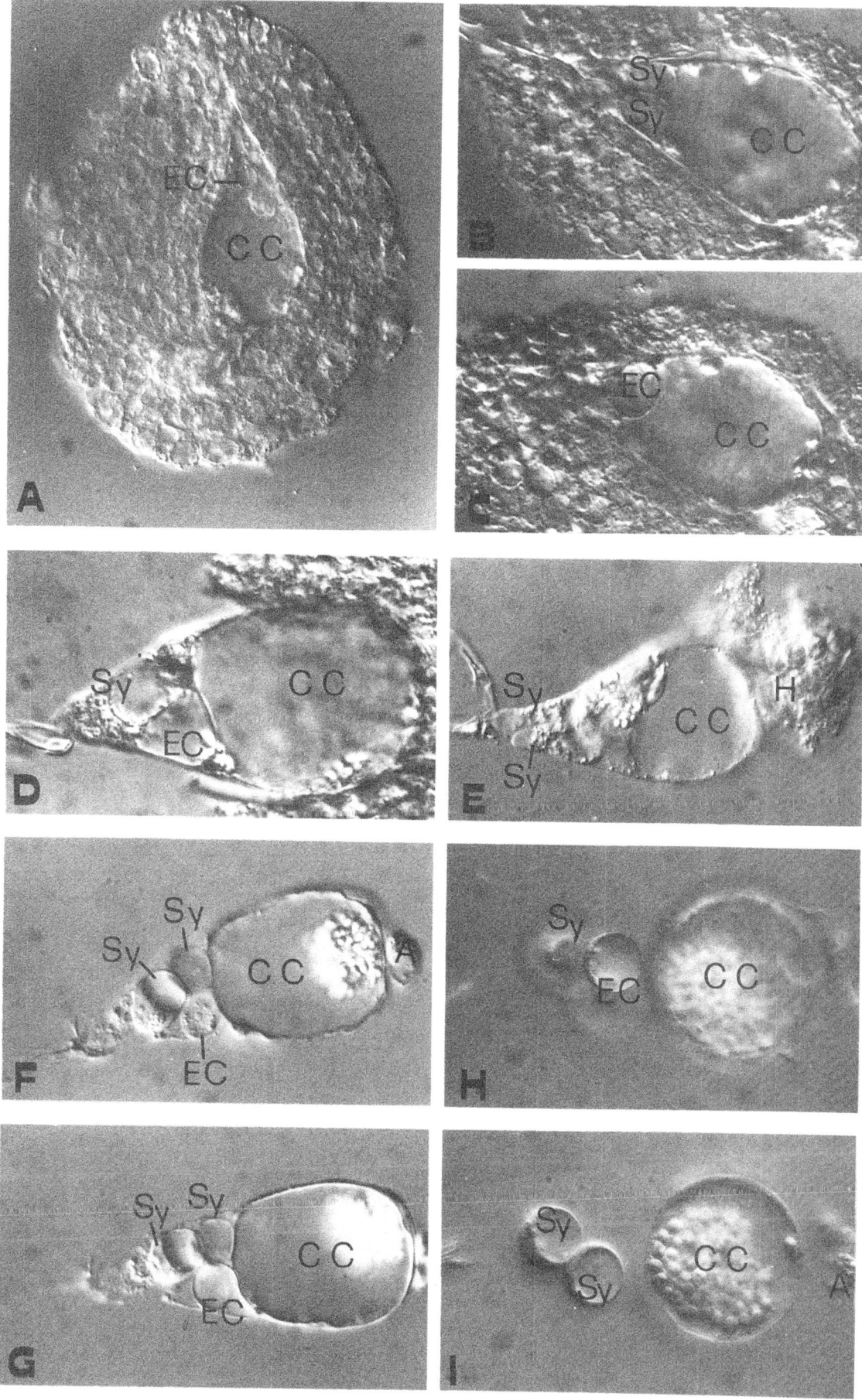

135

Plate 55A. Isolated embryo sac of *Nicotiana alata* Link & Otto, showing the central cell, one of the synergids and the egg cell. The cells are transforming into protoplasts, while the embryosac wall is still present. Video enhanced contrast differential interference contrast microscopy VEC-DIC, x 500.

Plate 55B. Isolated synergids and egg cell of *Nicotiana alata* Link & Otto. During the isolation procedure the cells have become spherical protoplasts. VEC-DIC, x 750.

Plate 55C. Isolated synergid of *Nicotiana alata* Link & Otto. VEC-DIC, x 750.

Plate 55D. Isolated embryo sac of *Plumbago zeylanica* L. The embryosac of *Plumbago* is composed of an egg cell and a central cell. In this species synergids are not formed. The microtubular cytoskeleton is visualised with FITC-anti tubulin staining and confocal laser scanning microscopy (CSLM), x 450.

Plate 55E. Isolated embryo sac of *Plumbago zeylanica* L. Actin filaments are visualized with rhodamine-phalloidine staining and CSLM, x 450.

(Plate 55A-E courtesy Bing Quan Huang, Norman and E S Pierson, Siena).

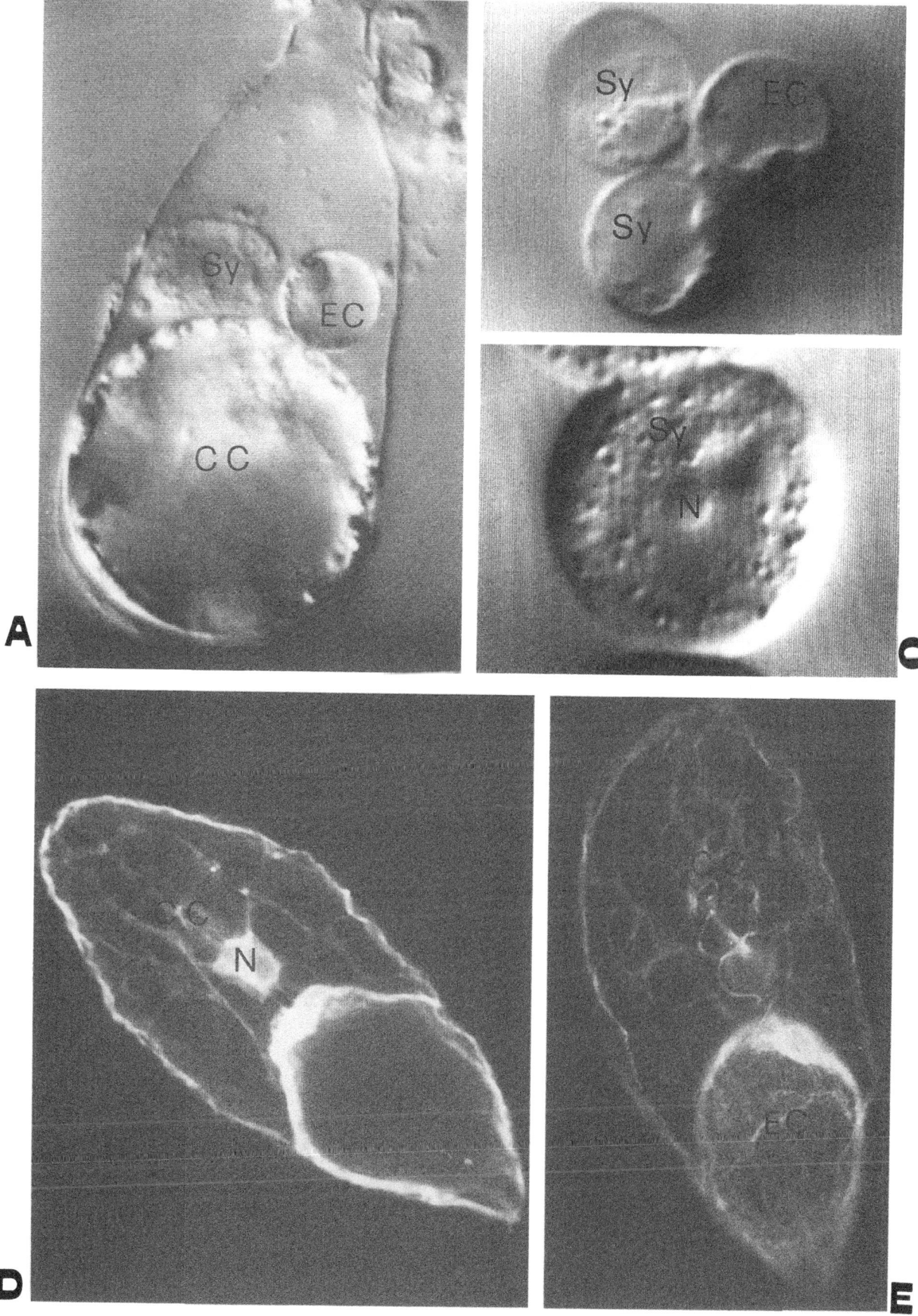

137

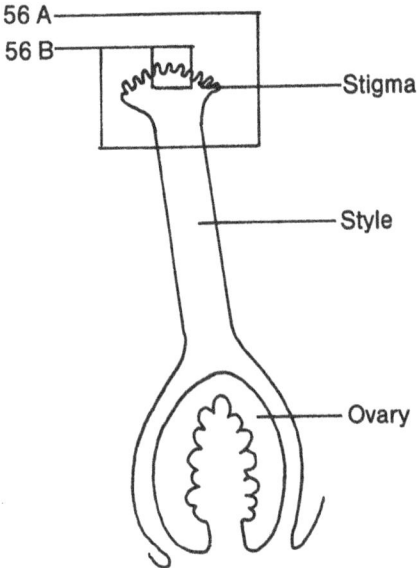

Plate 56A. Stigma of *Ipomoea purpurea* Roth. at the receptive stage. The stigma forms many lobes and is completely covered with short papillate extensions of the epidermis cells. During maturation there is no secretion of liquid exudate, and the stigma surface remains dry. x 560.

Plate 56B. Enlarged portion of the papillate surface of the receptive stigma of *Ipomoea purpurea* Roth. x 1,500

(Plates 56A,B courtesy F Ciampolini, Siena)

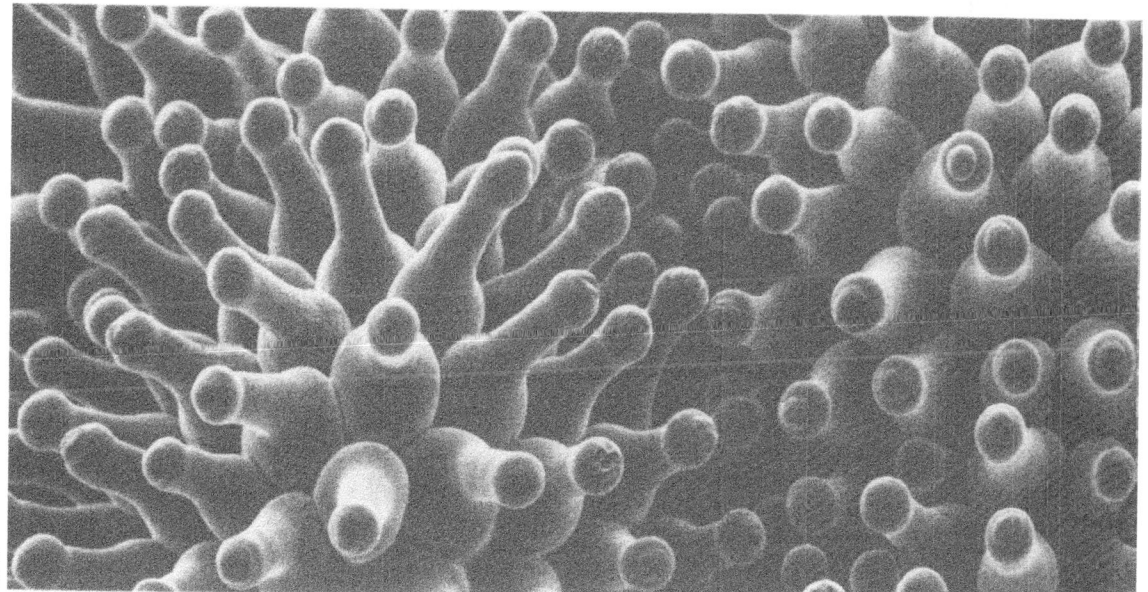

A

B

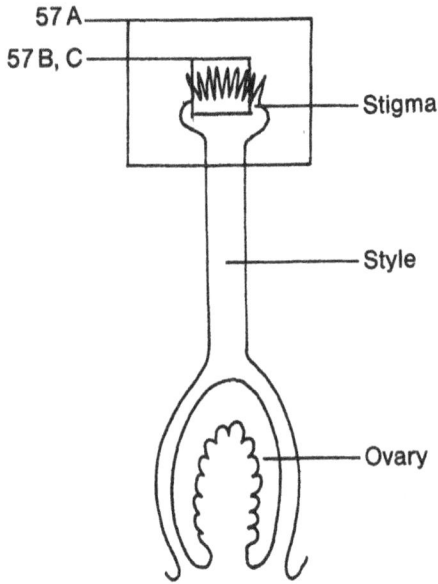

Plate 57A. Morphology of the dry stigma of *Hibiscus rosa-sinensis* L.. The stigma has already been pollinated. The pollen grains stick to the tip regions of the papillate stigma cells. x 70.

Plate 57B. Morphology of the dry stigma of *Arabidopsis thaliana* L. Heynh. The arrow points to a germinated pollen grain of which the pollen tube grows along the surface to the base of the stigma papillae. x 390.

Plate 57C. Portion of the dry stigma of *Hypericum calycinum* L. with adharing pollen grains. x 410.

(Plates 57A-C courtesy F Ciampolini, Siena).

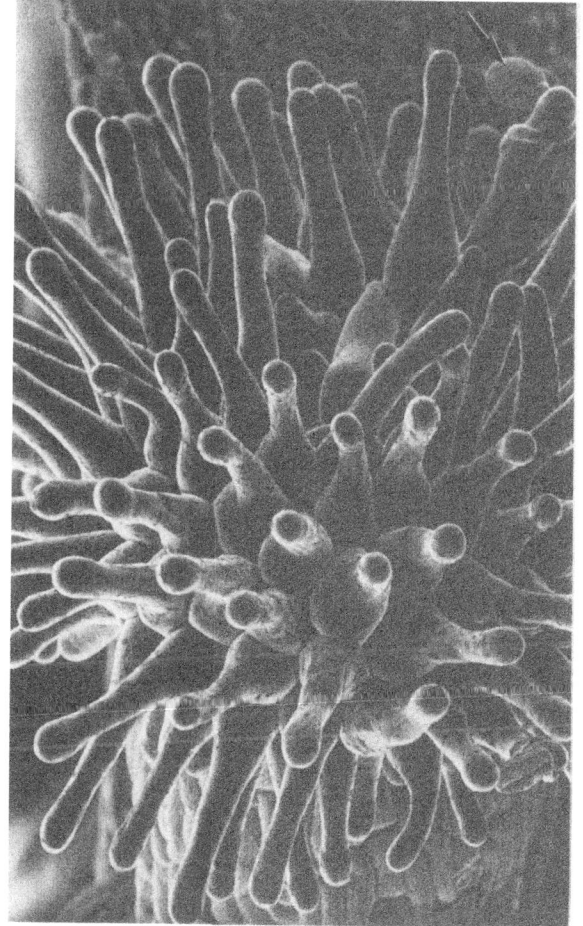

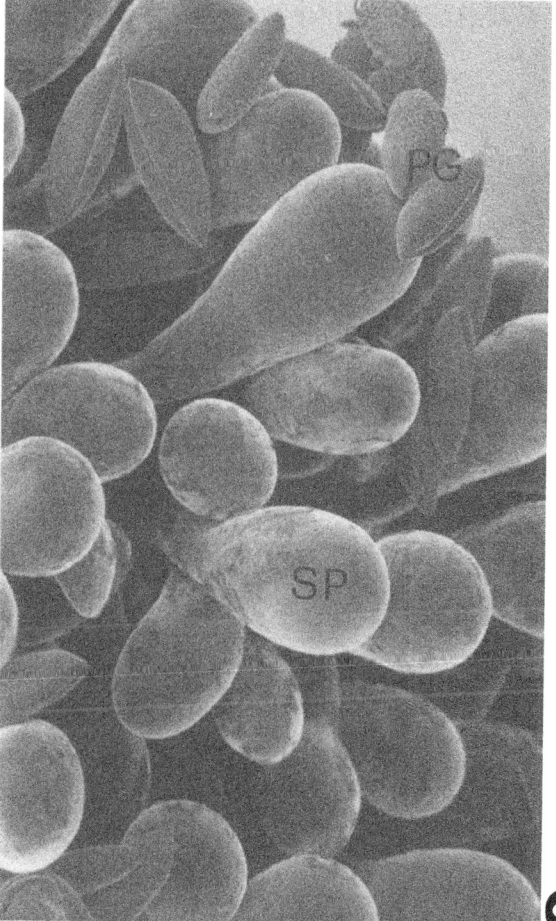

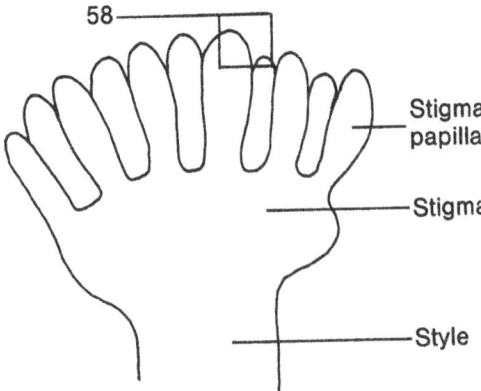

58

Stigma
papilla

Stigma

Style

Plate 58. Enlarged portion of a papilla cell of the *Hypericum calycinum* L. stigma. In the cytoplasm large accumulations of electron-dense material are formed. Electron-dense material is also deposited in the outer layers of the cell wall and below the cuticle. The cell wall is very thick, and multilayered with cellulosic microfibrils oriented differently in the various layers. The cell surface becomes covered by a pellicle, which breaks at anthesis, after which the surface becomes covered by stigmatic exudate. x 16,000.

(Plate 58 courtesy F Ciampolini, Siena).

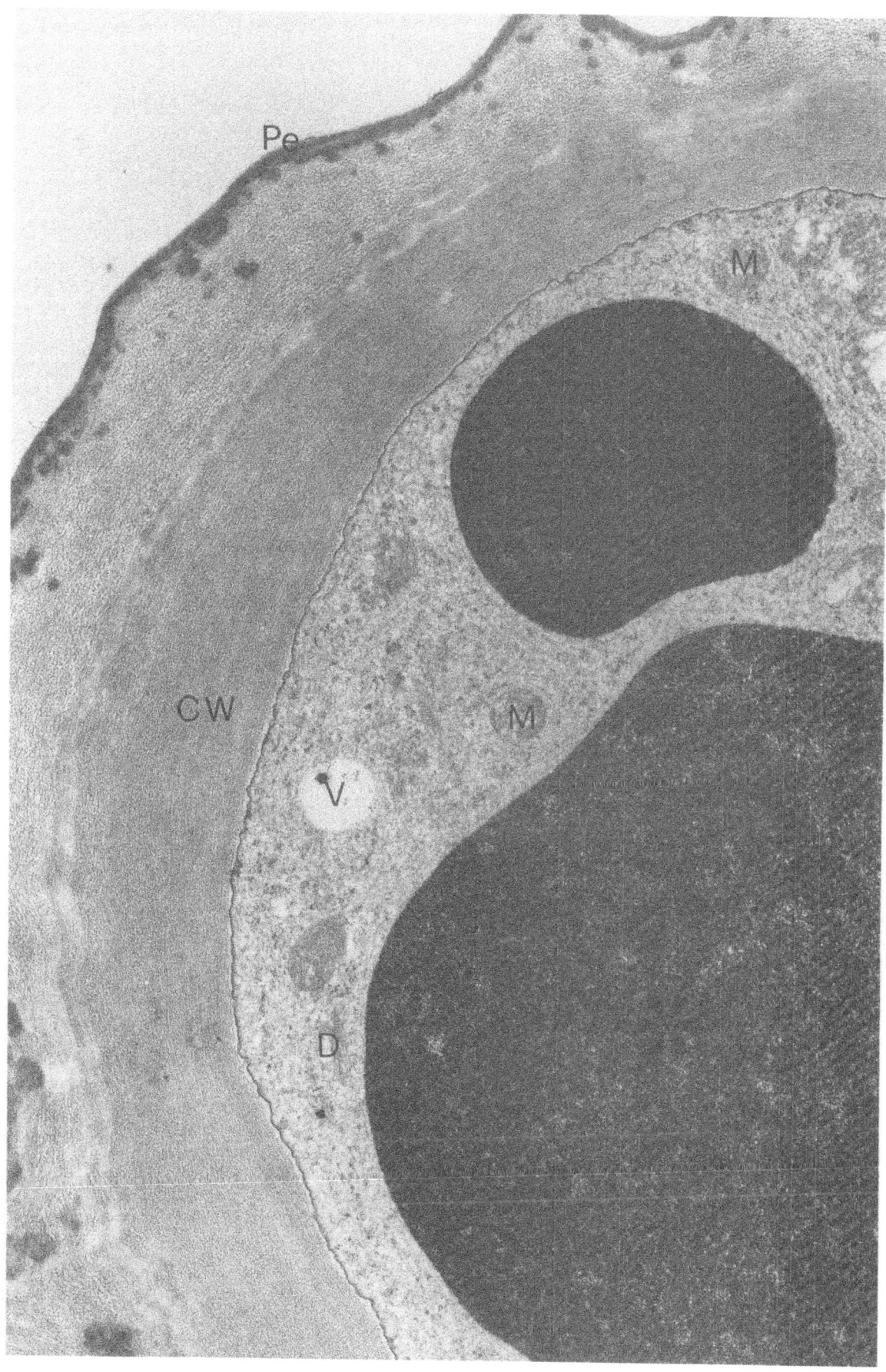

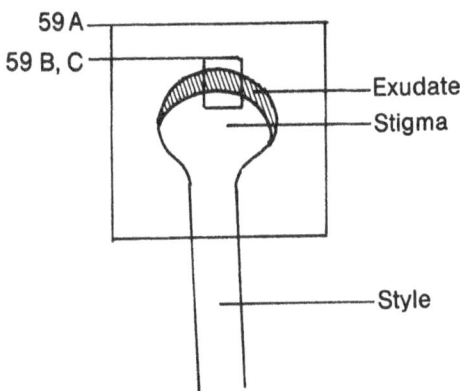

Plate 59A. Young style of *Nicotiana tabacum* L.. The style of *Nicotiana tabacum* ends in a broad bilobed stigma, which at maturity is completely covered by a large quantity of liquid exudate. At this young stage a large number of papillae are visible, and the secretion of exudate is still quite limited. x 40.

Plate 59B. Enlarged portion of the stigma surface of *Nicotiana tabacum* L. during maturation and exudate secretion. At many places of the surface, as well as onto the papillae (arrow) small droplets of exudate can be seen. The droplets increase in size and coalesce, eventually forming a thick layer. x 400.

Plate 59C. Styles of *Fragaria vesca* L. at the mature, receptive stage. Most styles are completely covered by a thick layer of exudate. x 140.

(Plates 59A-C courtesy F Ciampolini, Siena).

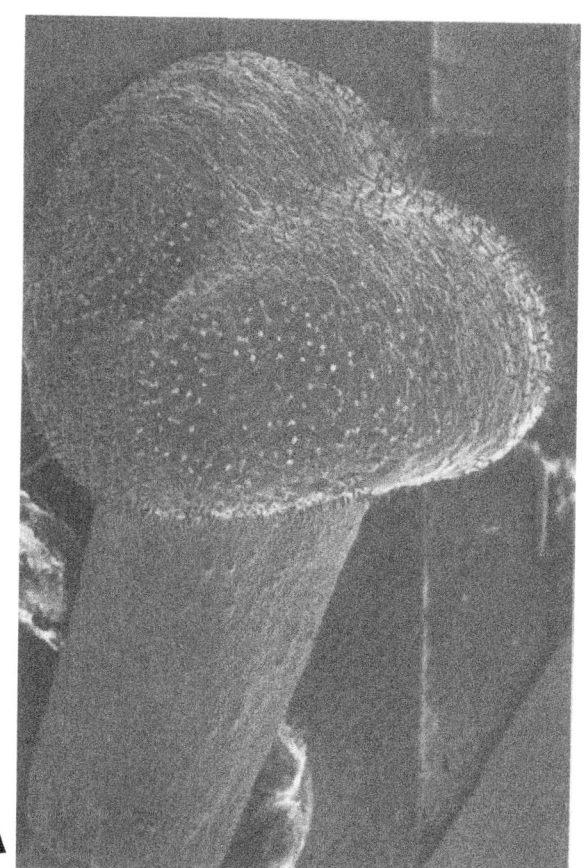

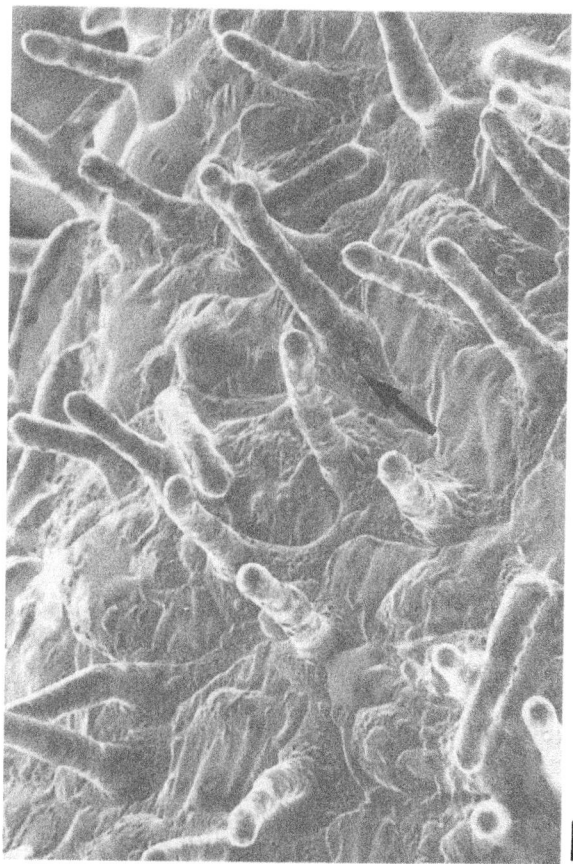

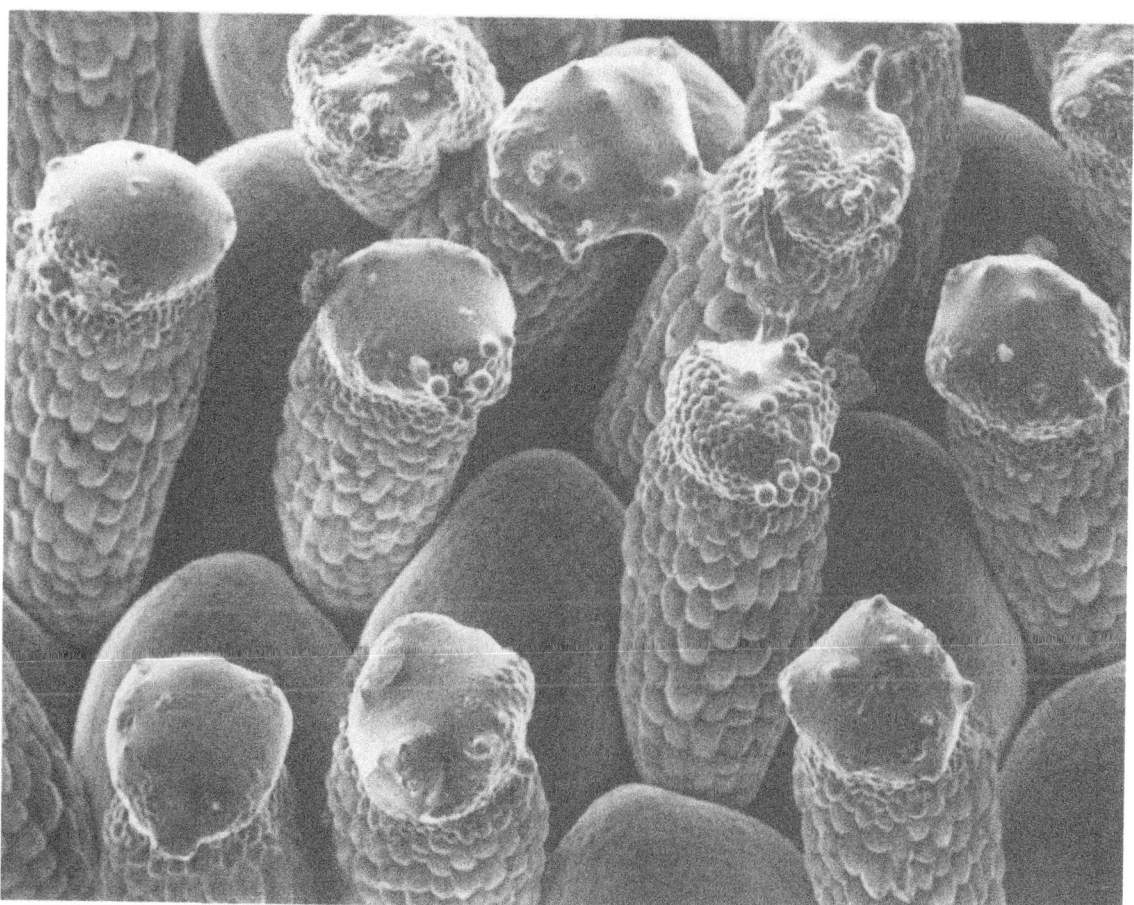

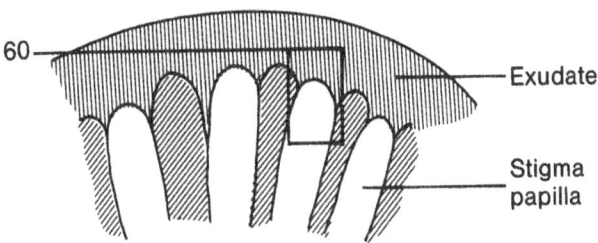

Plate 60. Transverse section through stigma papilla of the mature style of *Nicotiana tabacum* L.. The cytoplasm of the papilla is densely packed with both rough (arrow) and smooth endoplasmic reticulum (double arrow). The mitochondria contain large number of cristae. The plastids have only a reduced thylakoid system and contain starch. The papillae are completely surrounded by exudate which appears to be composed of electron dense droplets, imbedded in a granular matrix. x 19,000.

(Plate 60 courtesy F Ciampolini, Siena).

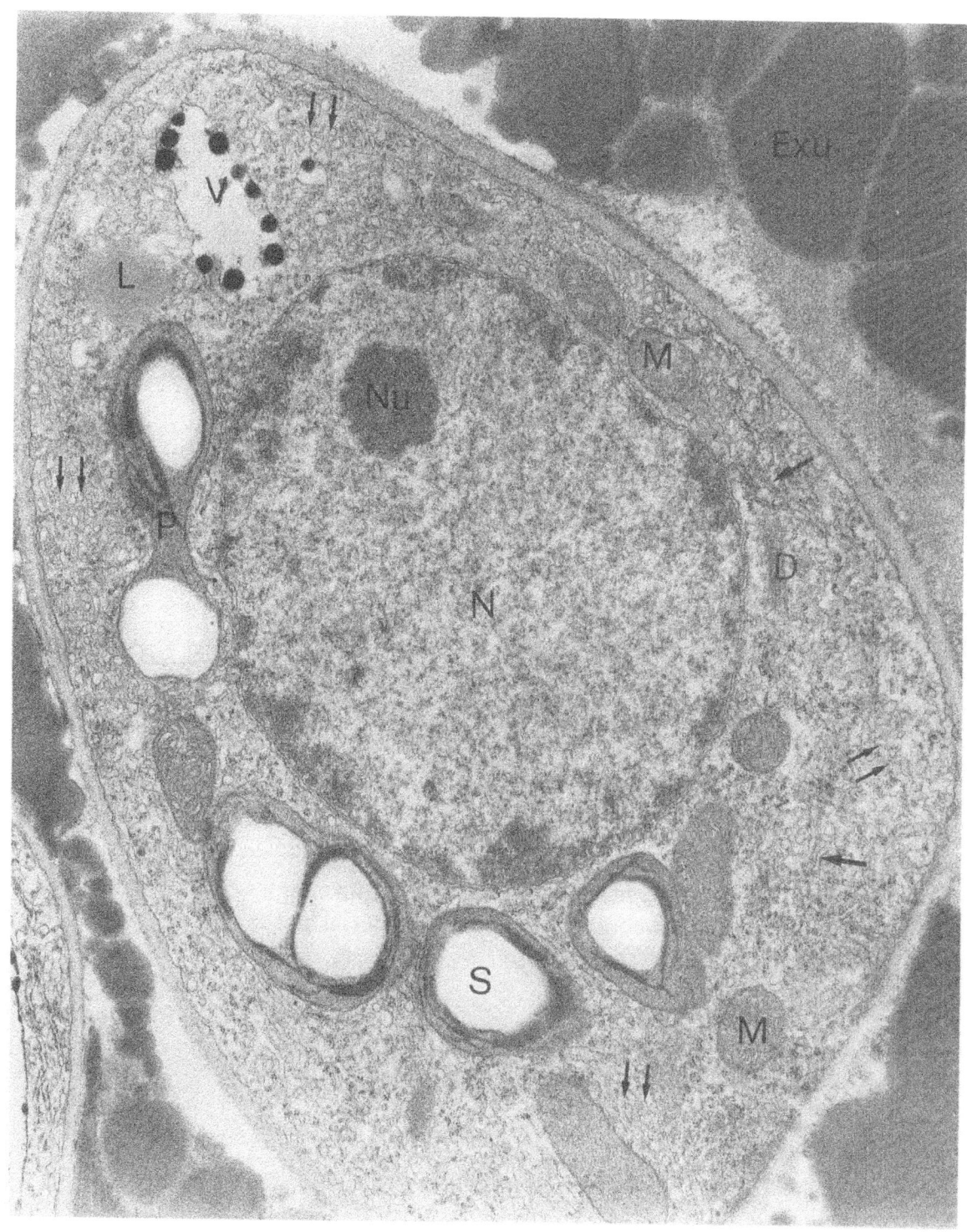

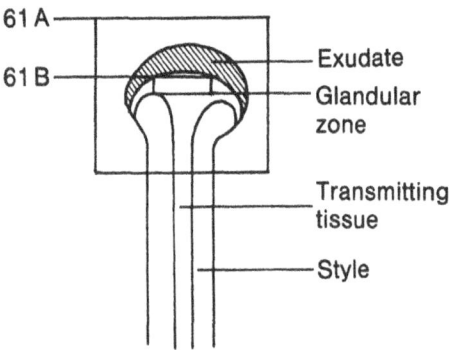

Plate 61A. Longitudinal section through the young pistil of *Nicotiana alata* Link & Otto, showing the transition zone between stigma and style. At the stigma surface the papillae are still visible (arrows), and exudate has not yet been secreted. Underneath the stigma surface there is a glandular zone which is continuous with the transmitting tissue of the style. The cells of the glandular zone are involved in the production and secretion of the stigmatic exudate. The peripheral zone of the style and the lower portion of the stigma are composed of non-glandular parenchyma cells. Within this tissue the vascular bundles can be seen. x 85.

Plate 61B. Transverse section through the style of *Nicotiana alata* Link & Otto, just below the stigma surface. This tissue is continuous with the stylar transmiting tissue, and composed of exudate forming cells. The intercellular spaces are filled with granular material and electron dense droplets (asterisks). x 18,000.

(Plate 61A reproduced by permission from M Cresti et al (1986) Amer J Bot 73: 1713-1722; Plate 61B original M Cresti).

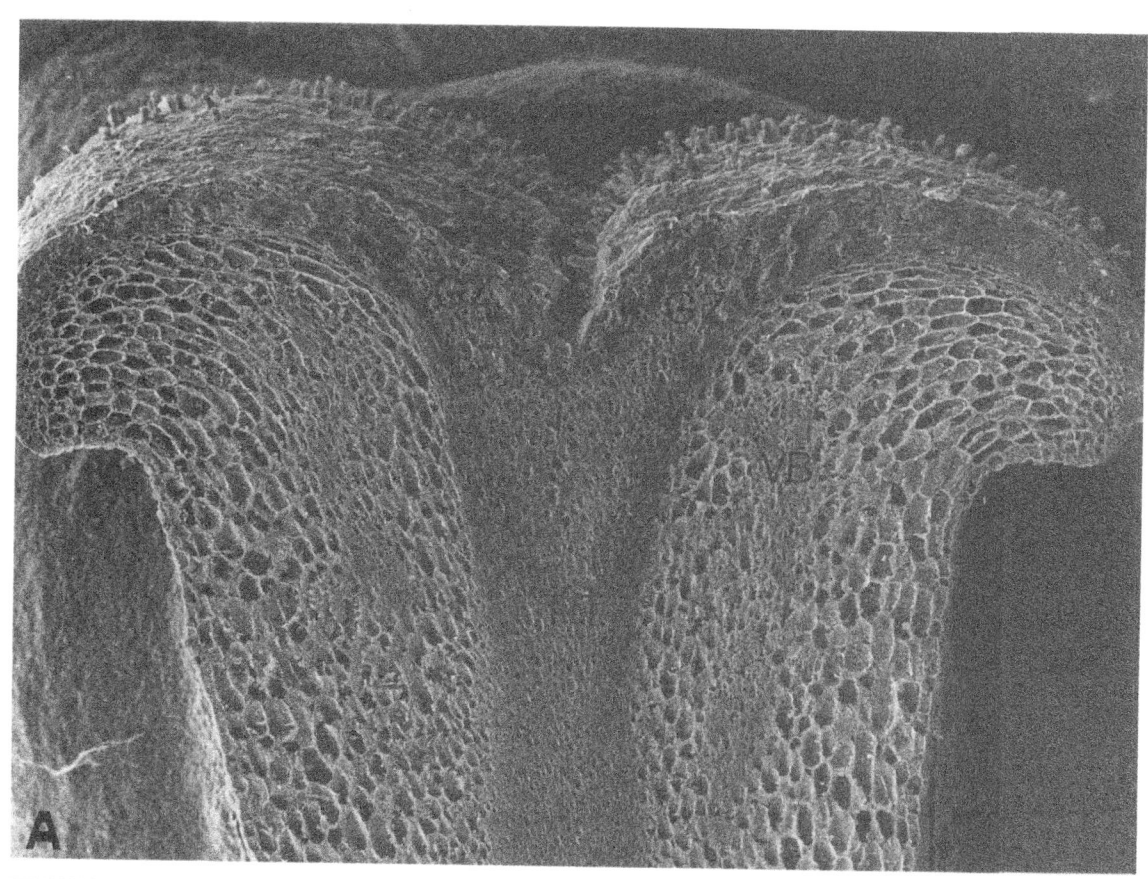

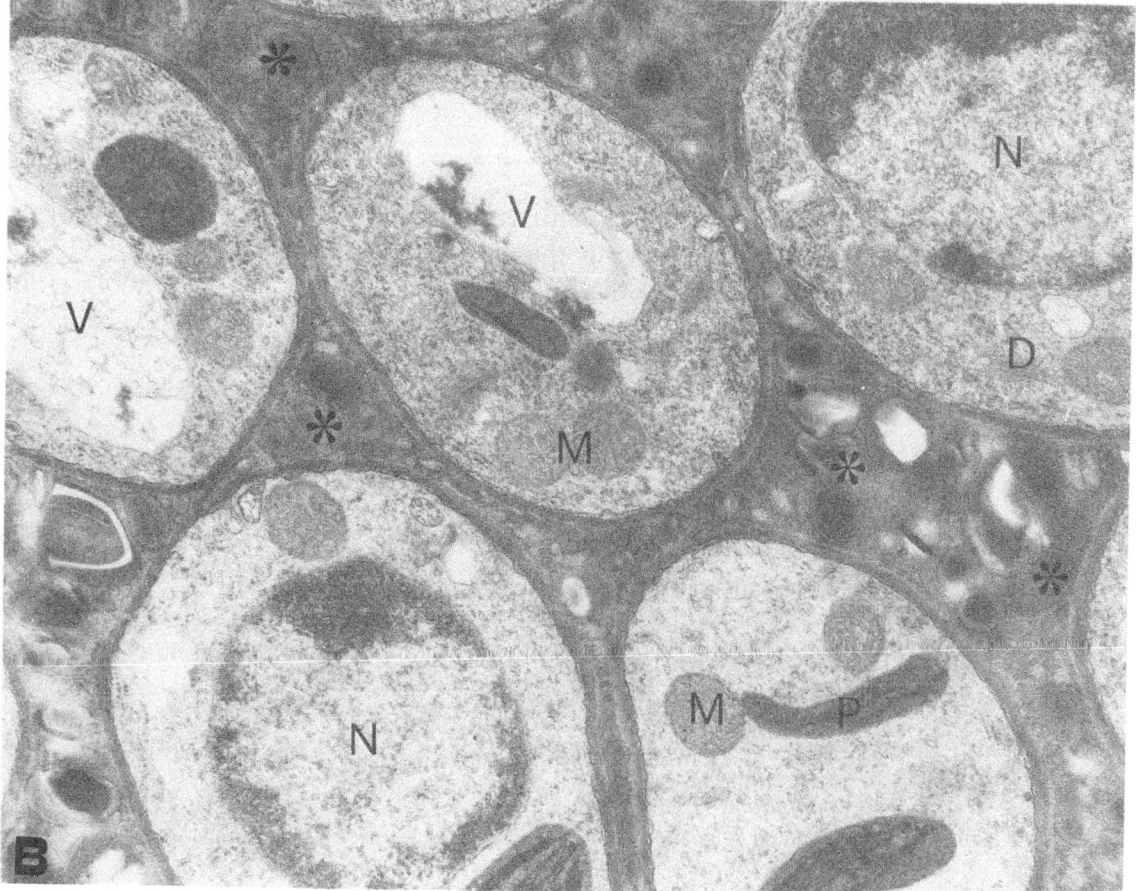

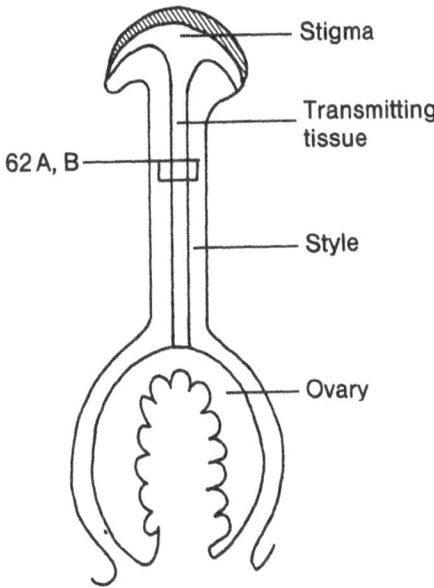

Plate 62A. Transverse section through a young style of *Lycopersicon peruvianum* (L.) Mill. During the stylar development the cells in the centre of the style form a solid column of transmitting tissue through which the pollen tubes will grow towards the ovary. The middle lamellae in between the cylindrical cells become strongly thickened. x 5,100.

Plate 62B. Detailed portion of the transmitting tissue of *Lycopersicon esculentum* (L.) Mill., showing the thickened and loosely organized middle lamellae. x 16,000.

(Original M Cresti).

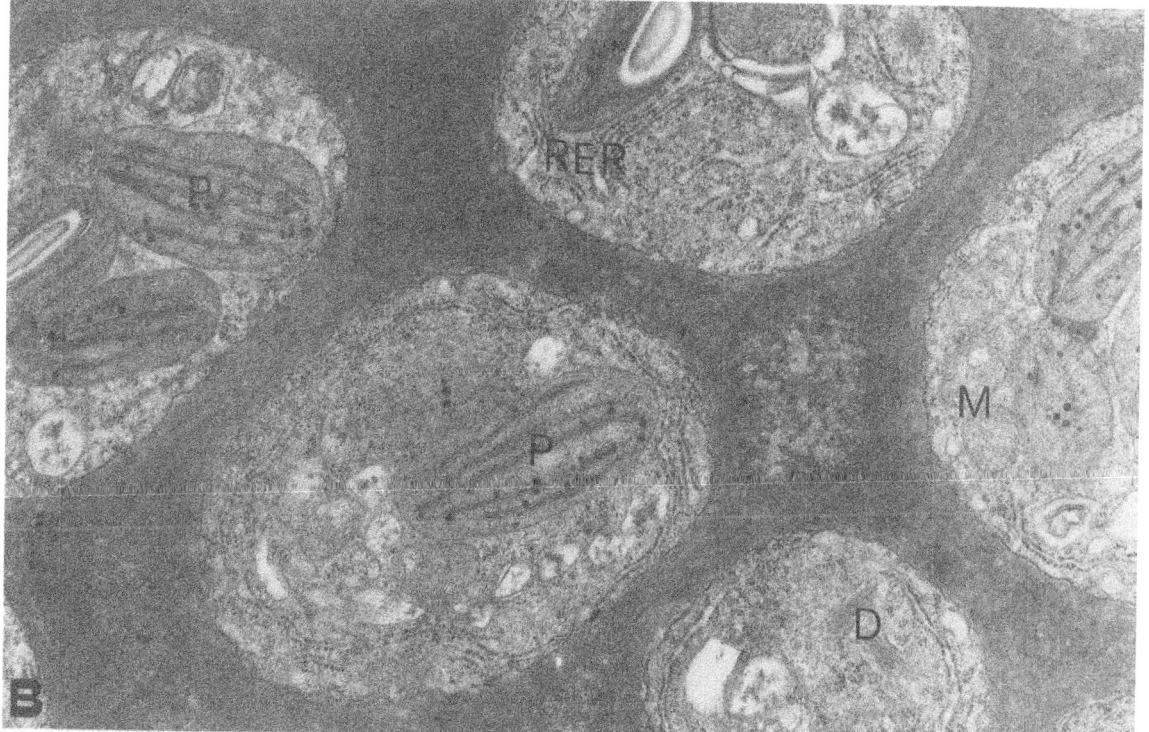

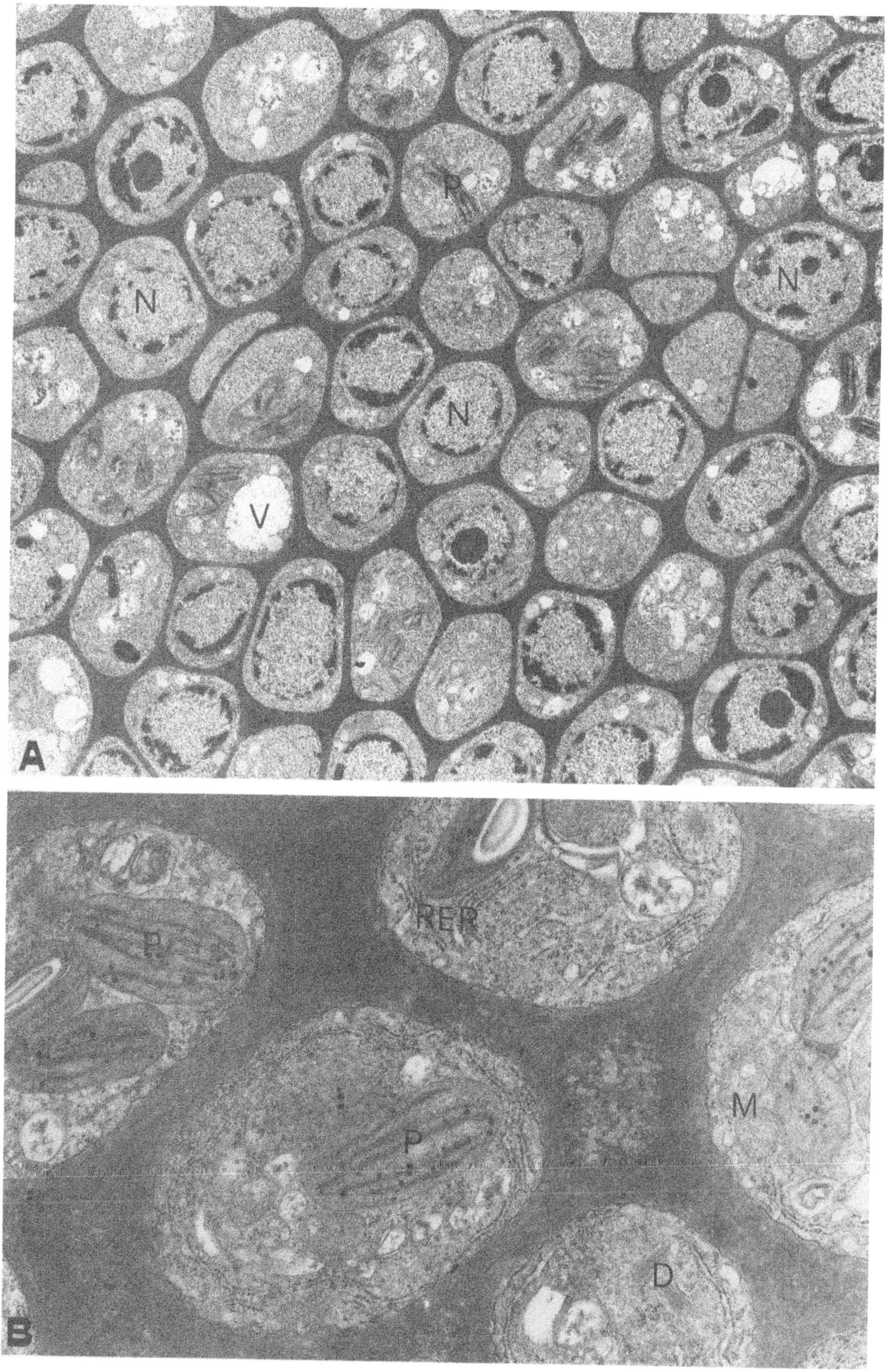

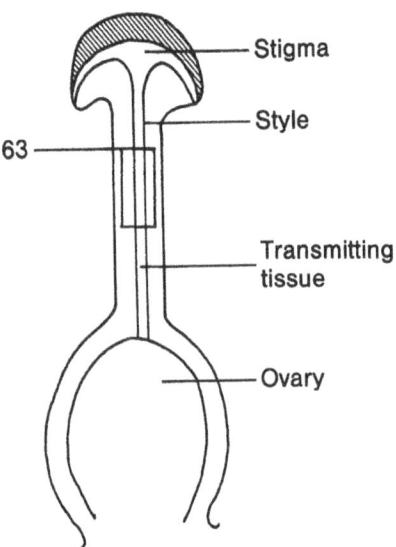

63 —

Stigma

Style

Transmitting
tissue

Ovary

Plate 63. Longitudinal section through the stylar transmitting tissue
of *Lycopersicon esculentum* (L.) Mill. It shows the elongated shape
of the cells which are interconnected by numerous plasmodesmata
in the transverse walls (arrowheads). The middle lamellae of the
longitudinal walls are thickened by the deposition of intercellular
substance. The transverse walls remain relatively thin. x 14,200.

(Original M Cresti).

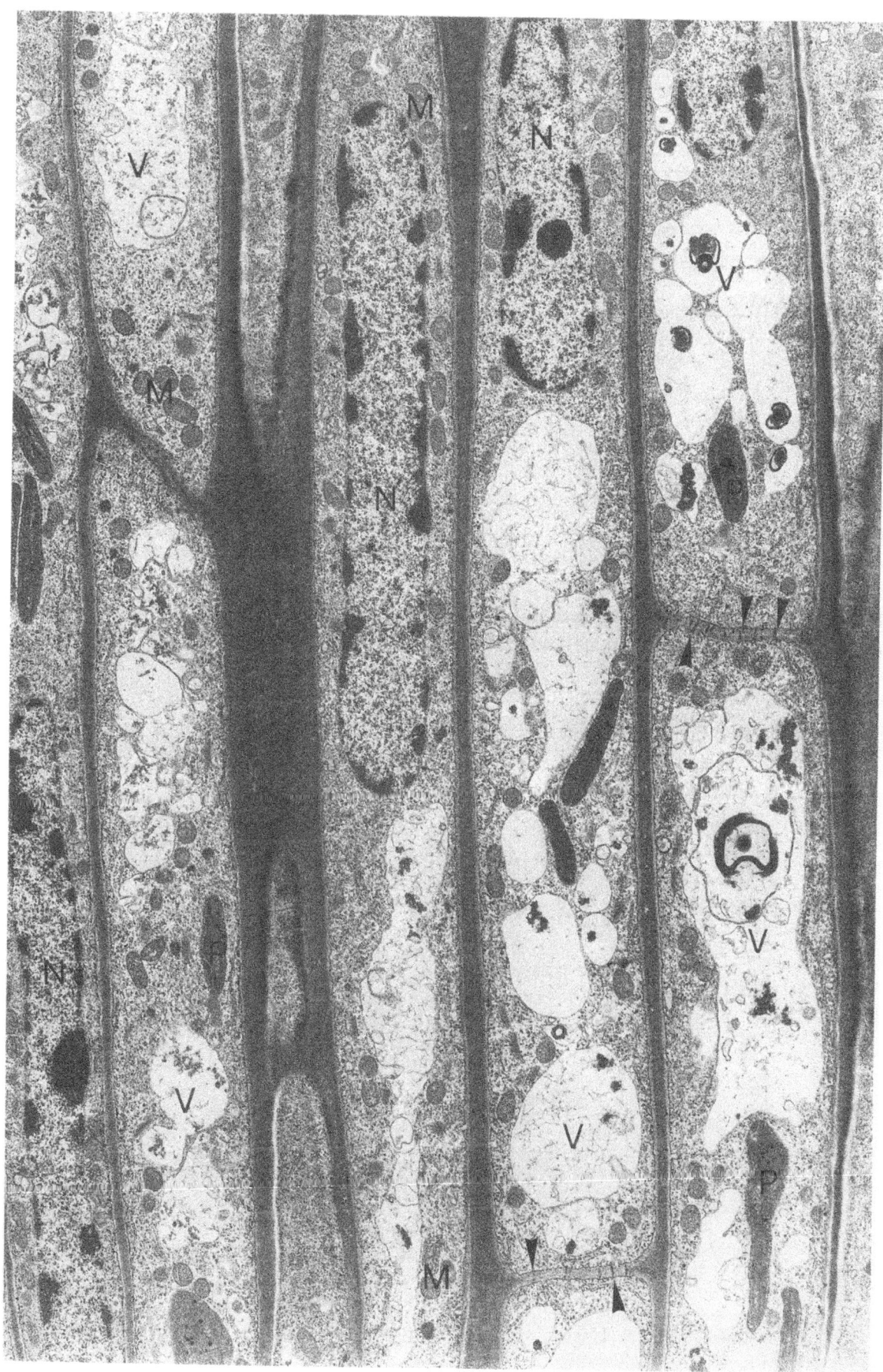

153

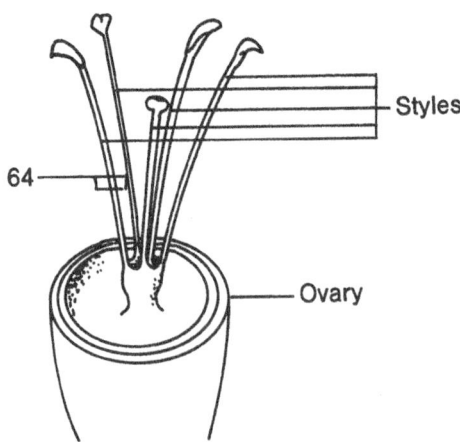

Plate 64. Transverse section through a mature style of *Malus communis* Poiret, showing the transmitting tissue. The transmitting tissue is composed of cylindrical cells and large intercellular spaces which are filled with intercellular substance. The intercellular substance is rather loosely organized and granular in structure, and clearly distinguishable from the cell walls. x 16,500.

(Plate 64 courtesy F Ciampolini, Siena).

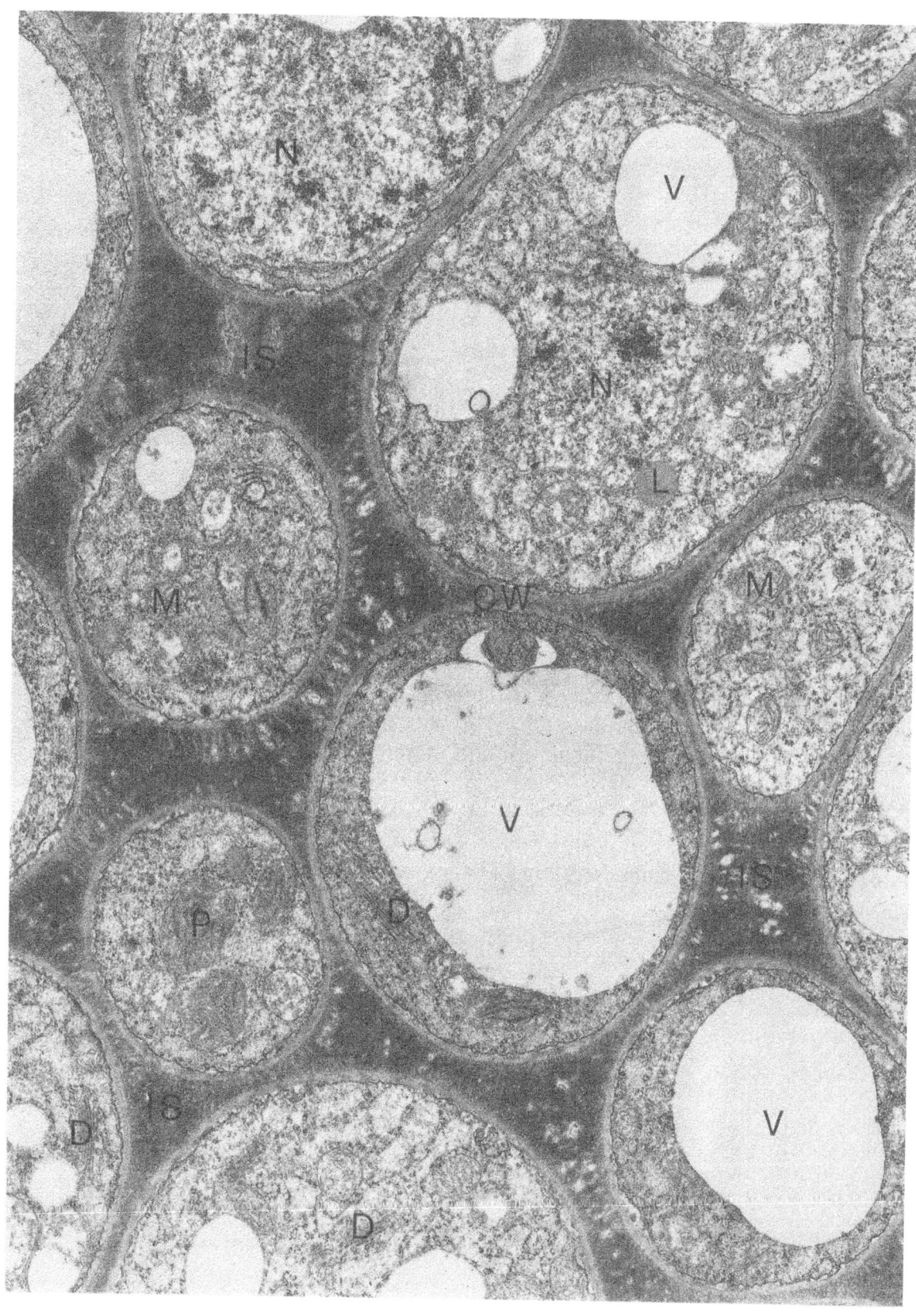

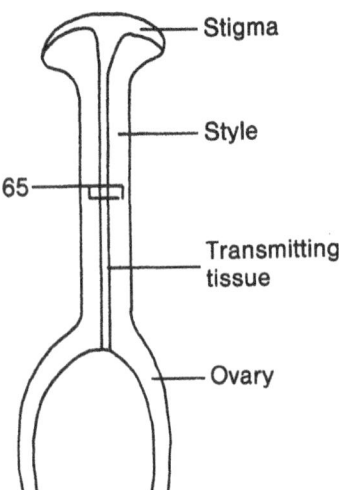

Plate 65. Transverse section through the style of *Olea europea* L., showing the stylar transmitting tissue. It is composed of elongated cells, of which the longitudinal, primary cell walls become very thick. The middle lamellae remain rather thin. The cells are interconnected by plasmodesmata in both the longitudinal (arrows) and transverse cell walls (asterisks). The thickened longitudinal, primary cell walls provide the pathway for the growing pollen tubes. The cells of the transmitting tissue contain starch and large accumulations of tannins. x 7,000.

(Plate 65 reproduced by permission from F Ciampolini et al (1983) Caryologia 36: 211-230).

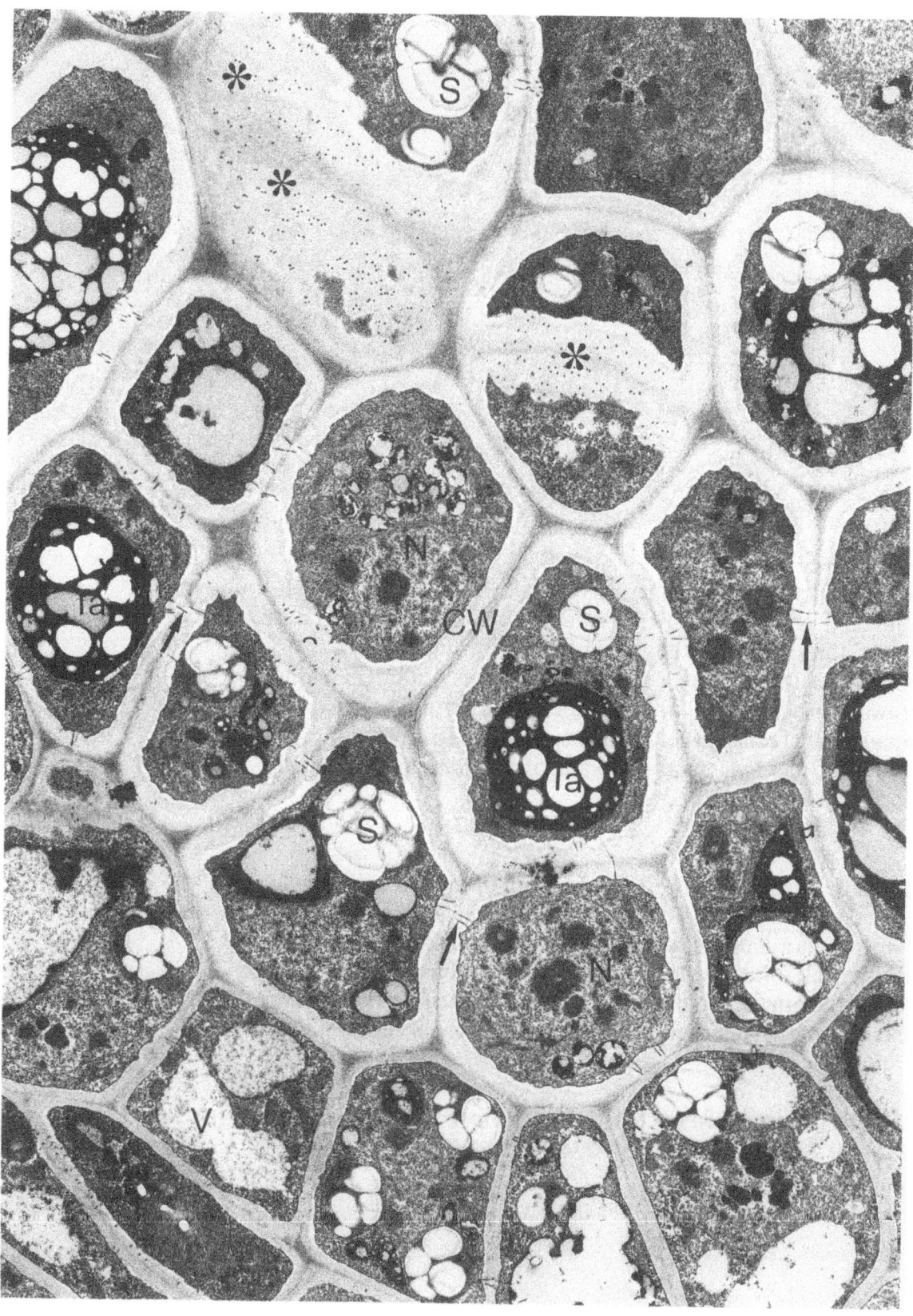

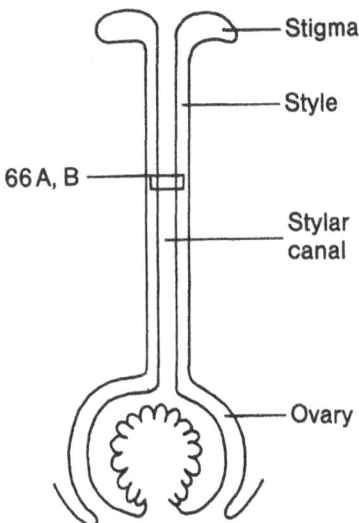

Plate 66A. Transverse section through the style of *Citrus limon* (L.) Burm. f., showing the narrow stylar canal, filled with electron dense exudate, through which the pollen tubes grow towards the ovary. The exudate (arrowheads), which accumulates under the cuticle (arrows) is produced by the glandular cells bordering the canal. The glandular canal cells are very rich in cytoplasm. During maturation of the style, the tangential cell walls of the canal cells gradually become thicker. x 5,100.

Plate 66B. Enlarged portion of a canal cell of the *Citrus limon* (L.) Burm. f. style, bordering the stylar canal. The cytoplasm contains numerous and prominent dictyosomes as well as many Golgi-vesicles (arrows). The dictyosomes are thought to be involved in both the secretion of the stylar canal substance, and the thickening of the inner tangential walls of the canal cells. x 48,600.

(Plate 66A reproduced by permission from F Ciampolini et al (1981) Pl Syst Evol 138: 263-274; Plate 66B courtesy of F Ciampolini, Siena).

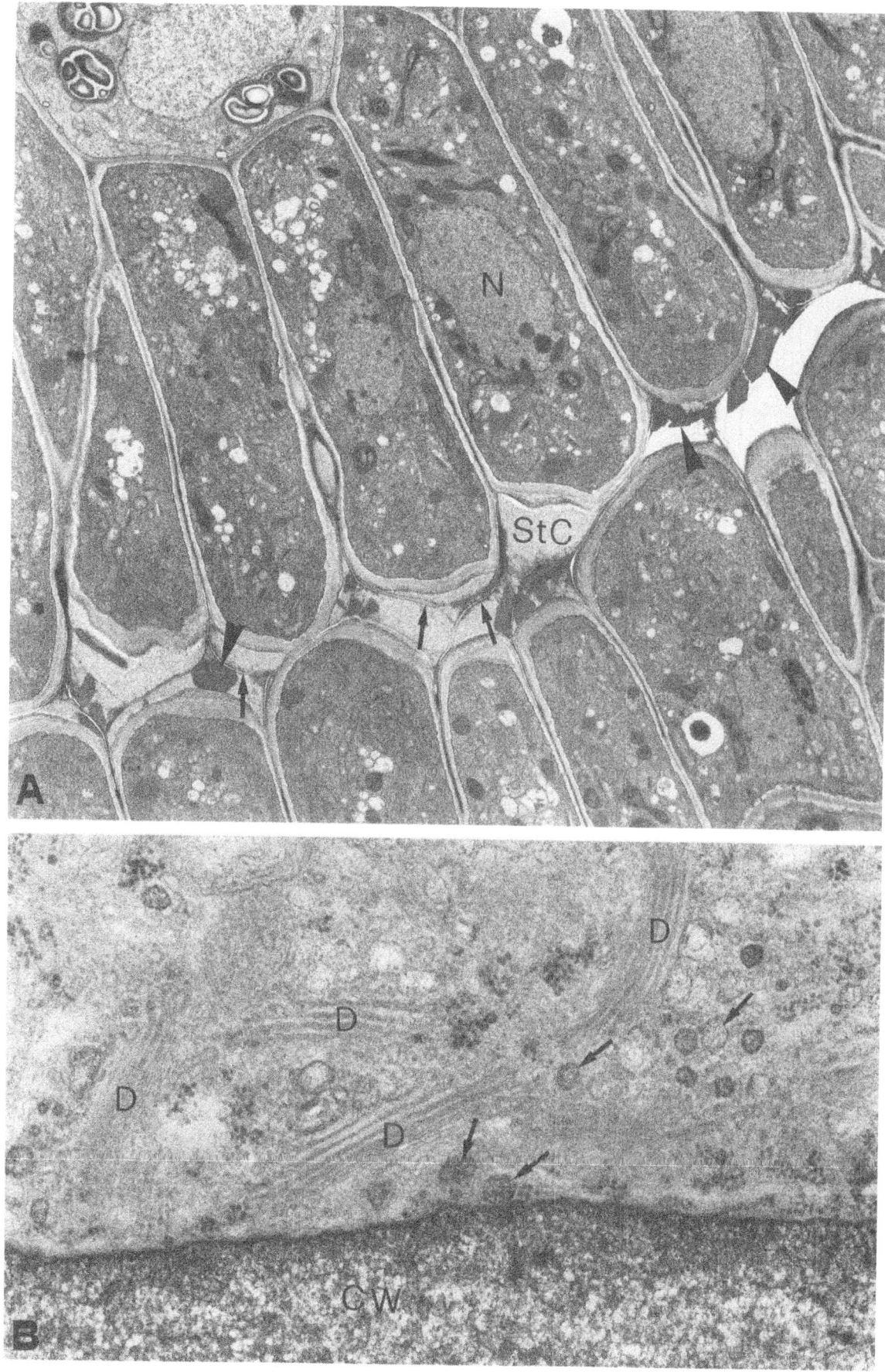

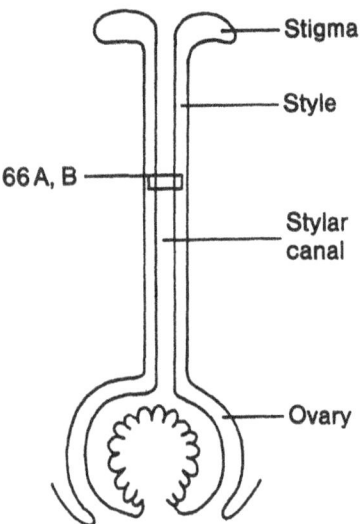

Plate 67A. Transverse section through the hollow style of *Lilium longiflorum* Thunb. The wide stylar canal is lined up with specialized, glandular cells, which produce and secrete an exudate into the canal. At maturity the wall of the canal is covered by a thick layer of exudate that provides the pathway for the pollen tubes. The stylar canal cells form a solid wall without interruptions or intercellular spaces and are very cytoplasmic rich, in contrast to the nieghbouring parenchyma cells. The tangential cell walls, bordering the stylar canal are much thicker then in other places and show many wall extensions. x 1,800.

Plate 67B. Enlarged view of a stylar canal cell of *Lilium longiflorum* Thunb., showing the complex organisation of the cytoplasm. The stylar canal is at the left side of the photograph. x 6,500.

(Original J Van Went).

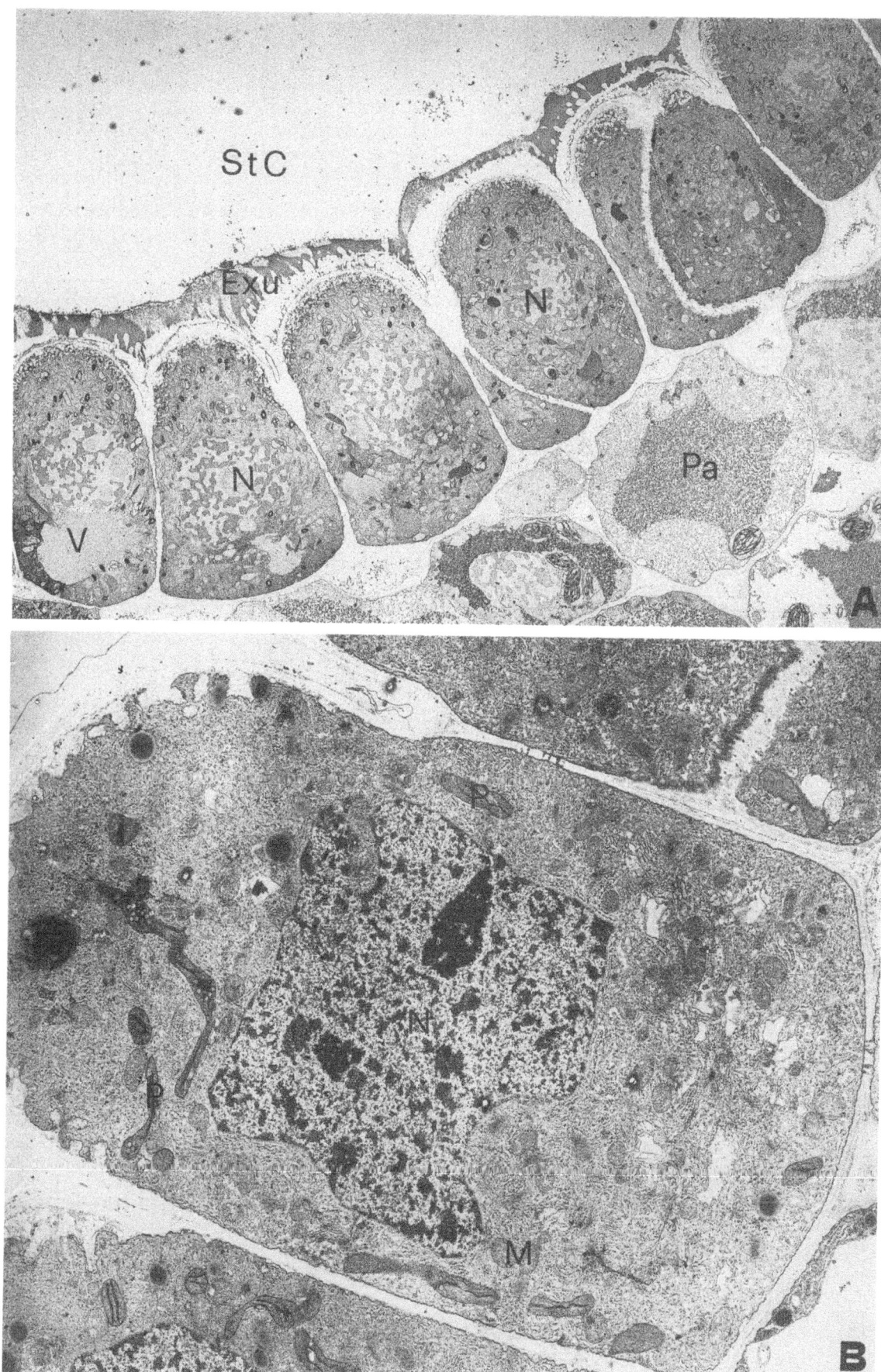

PART 3: PROGAMIC PHASE AND FERTILIZATION

PROGAMIC PHASE AND FERTILIZATION

Introduction

After the dehiscence of the anther, the pollen grains are either carried to the stigma by wind, animals, water, or directly by contact between the open anther and the stigma. The arrival of the pollen grains at the stigmatic surface marks the onset of the progamic phase, during which the male gametes are transported by the pollen tube to the female gametes. In angiosperms this results in double fertilization in which one sperm cell fuses with the egg cell, and the second fuses with the central cell.

The first interaction between pollen and stigma is attachment or capture of the pollen grains and this is accomplished by the sticky nature of the pollen surface, the stigmatic exudate, or the pellicle covering the stigma papillae. Attachment, in compatible pollinations, is immediately followed by uptake of water by the pollen grains through colloidal imbibition and endosmosis. This rehydration results in considerable swelling of the pollen grain, which in many species causes the release of proteins from the pollen wall. Such proteins are produced during pollen development by the tapetum cells, and are therefore of sporophytic origin. The sporophytic proteins are involved in recognition reactions between pollen and stigma. Some non-matching combinations lead to suppression of pollen germination, or failure of subsequent pollen tube penetration into the stigma. This phenomenon is called sporophytic incompatibility.

After rehydration, metabolic activity is resumed, which in turn leads to pollen germination. Some pollen grains germinate very rapidly and appear to be released from the anther fully prepared for immediate germination. In such pollen grains large numbers of Golgi-vesicles are present at maturity, containing the materials essential for rapid formation of pollen tubes. Other pollen types need considerable time to become activated and to germinate. In these types of pollen, activation is accompanied by considerable structural changes of the cytoplasm. Stacked RER cisternae become more dispersed, and the dictyosomes start to produce Golgi-vesicles containing cell wall precursors.

Pollen grain germination is marked by the formation of a pollen tube, emerging from one (or more) of the germination apertures. If supernumerous apertures are present, their outgrowth is generally blocked by deposition of callose. The wall of the emerging pollen tube is pecto-cellulosic in nature and continuous with the intine of the pollen grain . Elongation of the pollen tube occurs by strictly apical growth of the tube tip. After recognition and acceptance, the elongating and compatible pollen tube penetrates the stigma surface and underlying stigmatic tissues, or enters into the stylar canal. The pathway of the pollen tube shows great diversity, related to the specific structure of the stigma and style. Penetration of the stigma can require the excretion of cutinase and cell wall degrading enzymes by the pollen tube to breakdown the cuticle and cell wall components of the stigma in those plant types with dry stigma and massive styles. In plants with wet stigmas and massive styles, the pollen tubes reach the stylar transmitting tissue via the pathway formed by the exudate upon the stigma and the intercellular exudate in between the stigmatoid cells underneath the stigma surface. In plants with hollow styles, the pollen tubes grow along the surface of the stigma and the surface of the stylar canal cells. These surfaces can either be dry or covered by exudate, produced by glandular cells.

The route of the pollen tubes in a massive style depends on the structure of the transmitting tissue. In many species a pathway is provided by the presence of an intercellular substance or thickened middle-lamellae. In other species the pollen tubes grow through the longitudinal, thickened cell walls. They can, however, even penetrate the stylar cells and continue their growth through the cell lumina. During pollen tube growth in the style there are intensive interactions between the growing pollen tubes and the surrounding stylar tissue. These interactions comprise the enzymatic degradation of stylar components by the pollen tubes and their use as nutrition, as well as style-pollen tube recognition reactions. In many species, a gametophytic incompatibility system operates in the style, which only allows the passage of compatible pollen tubes. The growth of incompatible pollen tubes usually is arrested at a specific site in the style.

In the growing pollen tube, four zones with characteristic ultrastructural organization are found. The extreme apical region is completely filled with cytoplasm which contains many Golgi-vesicles. These supply the cell wall components and plasma membrane needed for the tip growth of the pollen tube. At

166

the apex of the tube the wall is very thin and pecto-cellulosic. The cell wall components needed for tube elongation are excreted by exocytosis at the extreme tube tip. Next to the apical region there is a zone which is also cytoplasm-rich, but in which many other organelles, especially mitochondria, dictyosomes and endoplasmic reticulum, are found in addition to Golgi-vesicles. This subapical region is metabolically very active, contributing to the synthesis of wall precursors and cytoplasmic proteins. In this region an additional wall layer is formed which contains callose and, has increasing rigidity and stability. The subapical cytoplasm-rich region gradually merges into the next zone, a vacuolated region containing the vegetative nucleus and the generative cell or sperm cells. Cytoplasmic streaming and the movement of the vegetative nucleus and the generative cell or sperm cells are observed in this region. It is probable that this movement is generated by microfilaments. In the peripheral region of the pollen tube a cytoskeleton composed of longitudinally oriented, cortical microtubules is present. The above, living portion of the pollen tube remains restricted in length. At regular intervals during tube elongation, callose plugs are formed, which separate the living and growing part of the tube from the rest of the pollen tube.

In bicellular pollen types, sperm cell formation takes place during pollen tube growth. During nuclear and cell division of the generative cell, the organization of its microtubular cytoskeleton strongly changes. Cytokinesis, generally occurs by the combined action of cell plate formation and cell constriction. The sperm cells remain positioned close to each other and the vegetative nucleus, forming the male germ unit.

In species with a gametophytic incompatibility system, the recognition mechanism that leads to blocking of pollen tube growth is unknown. The incompatible pollen tube usually has a very thick callosic wall, and the tube appears to burst at its tip.

The compatible pollen tube that reaches the ovary, grows along the surface of the placenta towards the ovules. In some species additional specific structures, such as the obturator or papillate placenta cells, provide a well-defined pathway which guides the pollen tube to the apex of the ovules. After reaching the ovule, the pollen tube normally reaches the embryo sac by entering the micropyle.

Clearly, the entire growth of the pollen tube from stigma to ovule is actively directed although the nature and regulation of the system are only poorly understood. Many hypotheses have been postulated and tested, ranging from chemotropism to mechanically and electro-physical directed growth systems. The same holds for the final stage of directed pollen tube growth to the embryo sac. It is generally assumed that this final stage of tube growth is directed chemotropically by substances produced by the ovule and secreted through the micropyle. The synergids of the embryo sac are supposed to produce these chemotropic substances, and to excrete them into the filiform apparatus. From the filiform apparatus the substances should leach out into the micropyle, or the adjacent nucellus cells, creating a gradient which guides the pollen tube.

In species with tenuinucellate ovules, the mature embryo sac is not surrounded by nucellus tissue, and the pollen tube reaches directly the embryo sac after passage of the micropyle. In species with crassinucellate ovules the mature embryo sac is surrounded by nucellus tissue, and the pollen tube has to enter the nucellus tissue, after it has passed the micropyle in order to reach the embryo sac. In these cases, it has been found that more then one pollen tube can reach the nucellus, but only one pollen tube actually penetrates. Apparently, a recognition system also operates here, leading to the rejection of supernumerous pollen tubes.

Penetration of the pollen tube into the embryo sac starts with the growth of the tube into and through the filiform apparatus, after which it enters one of the synergids. In species in which one of the synergids has degenerated before the arrival of the pollen tube, it is invariably into this degenerated synergid that the pollen tube enters. As soon as the pollen tube has reached the synergid cytoplasm, its growth ceases, and the tube tip bursts or a terminal pore is formed. Via this aperture a considerable portion of the tube content, including the vegetative nucleus and the two sperm cells is injected into the synergid. This causes drastic swelling of the penetrated synergid, and degenerative changes of both tube and synergid cytoplasm. The plasma membrane of the penetrated synergid degenerates or ruptures, and degenerating cytoplasm can be seen penetrating in between the adjacent cells. Also the vegetative plasma membranes around the sperm cells degenerate, but both sperm cells themselves remain unaffected. The ultrastructure of the other cells of the embryo sac is not affected by the penetration of the pollen tube and the degeneration of the penetrated synergid.

The two sperm cells are thought to be transported or to move in an unknown way via the penetrated synergid to the base of the egg cell and the central cell. Subsequently, one sperm cell fuses with the egg cell, and the second sperm cell fuses with the central cell. Fusion of the male and female gametes is supposed to result from local contact and fusion of their respective plasma membranes. The fusion of male and female plasma membranes results in the formation of a passage between the gametes through which the sperm cell nucleus and cytoplasm can enter the female gamete. In this proposed mechanism of fusion there is no requirement for the formation of pores in the female gamete plasma membrane to facilitate the transfer of male material. It also provides a mechanism for preventing the entrance of degenerated synergid or pollen tube cytoplasm into the female gametes.

The next step in the process of double fertilization is the fusion of the sperm cell nucleus with the egg cell nucleus, and the fusion of the second sperm cell nucleus with the (fused) polar nuclei. The nuclear fusions start with local contacts and fusions of subsequently the outer nuclear membranes and the inner nuclear membranes. In this way nucleoplasmic bridges are formed between the fusing nuclei, which enlarge and coalesce. Finally the content of the male and female nuclei intermingle, which marks the completion of fertilization.

Soon after fertilization is achieved, the nucleus of the fertilized central cell starts to divide. Usually, the initial nuclear divisions are not accompanied by cytokinesis, which results in the formation of nucleate endosperm. The nucleate endosperm (or part of it) can become cellular at a later stage of development. In many species the endosperm accumulates and stores reserve substances, which serve as food supply for the growing embryo.

In general, embryo formation starts later then endosperm formation. Before the zygote divides, its shape and cytoplasmic content drastically change, and a polar distribution of the cell organelles is achieved. The first division results in a larger, vacuolated micropylar cell, and a smaller, cytoplasm-rich chalazal cell. From the smaller, cytoplasm-rich chalazal cell the embryo proper derives, while the larger, more vacuolated cell develops into the suspensor. During embryo and endosperm formation the integuments develop into the seed coats, which protect the newly formed sporophytic plant during its dispersal.

Recommended literature

Cresti M, Ciampolini F, Sarfatti G (1980) Ultrastructural investigations on *Lycopersicum peruvianum* pollen activation and pollen tube organization after self and cross-pollination. Planta 150: 211-217

Cresti M, Ciampolini F Mulcahy D L M, Mulcahy G (1985) Ultrastructure of *Nicotiana alata* pollen its germination and early tube formation. Amer J Bot 72: 719-727

Cresti M, Dallai R (eds) (1986) Biology of reproduction and cell motility in plants and animals. University of Siena, Siena

Cresti M, Pacini E, Ciampolini F, Sarfatti G (1977) Germination and early tube development in vitro of *Lycopersicum peruvianum* pollen: ultrastructural features. Planta 136: 239-247

Heslop-Harrison J (1972) Sexuality of angiosperms. In: Steward F C (ed) Plant Physiol, Vol VI C. Academic Press, New York-London, pp 133-209

Heslop-Harrison J (1987) Pollen germination and pollen tube growth. Int Rev Cytol 107: 1-78

Knox R B (1984) Pollen-pistil interactions. In: Linskens H F, Heslop-Harrison J (eds) Cellular interactions. Encycl Plant Physiol, Vol 17. Springer, Berlin-Heidelberg-New York-Tokyo, pp 508-608

Linskens H F (ed) (1964) Pollen physiology and fertilization. North-Holland, Amsterdam

Nettancourt D de (1977) Incompatibility in angiosperm. Springer, Berlin-Heidelberg-New York

Went J L van, Willemse M T M (1984). Fertilization. In: Johri B M (ed). Embryology of angiosperms. Springer, Berlin- Heidelberg-New York-Tokyo. pp. 273-318

Plate 68A. Pollinated pistil of *Hibiscus rosa-sinensis* L.. The stigma of *Hibiscus* is of the dry type and is covered by many long, needle-like papillae, to which the deposited pollen grains adhere. x 1,800.

Plate 68B. Pollinated pistil of *Ipomoea purpurea* Roth. The large pollen grains land on the papillate dry stigma and are in contact with a number of papillae. x 6,500.

(Plates 68A, B courtesy F Ciampolini, Siena).

A

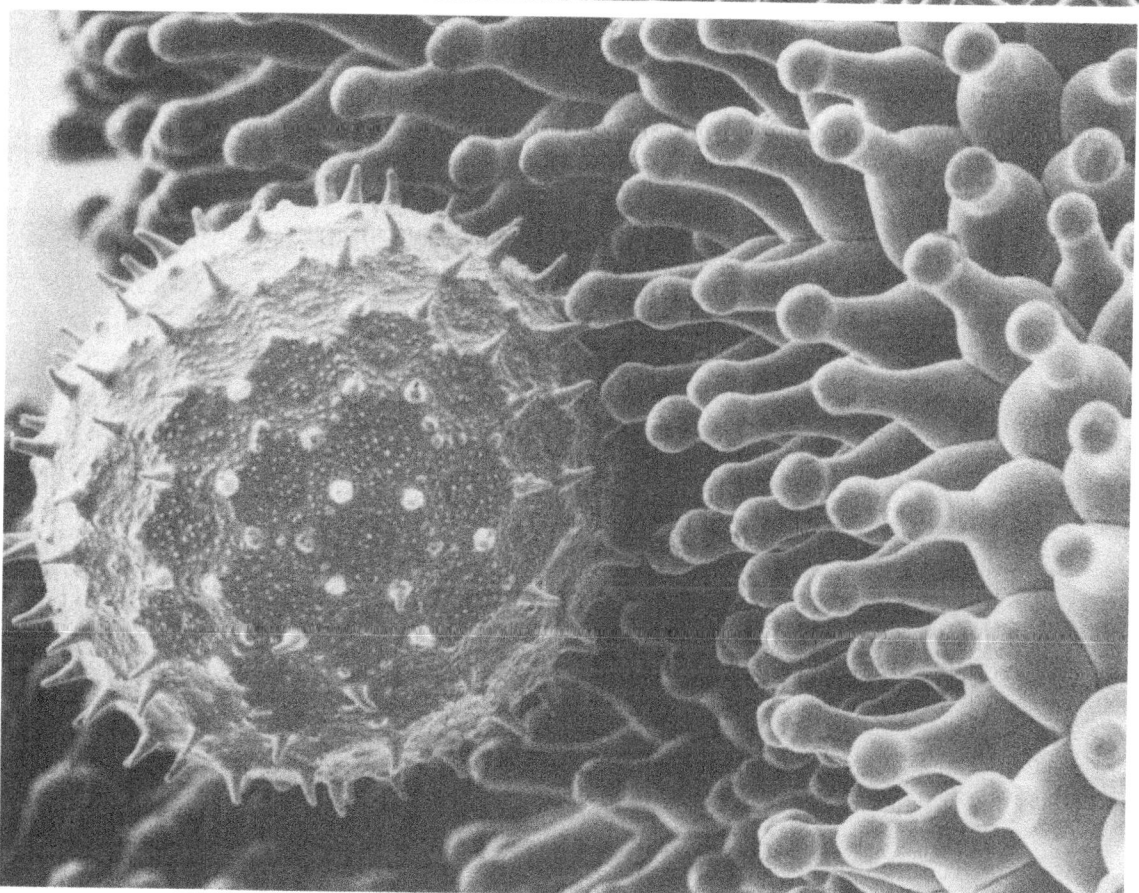

B

173

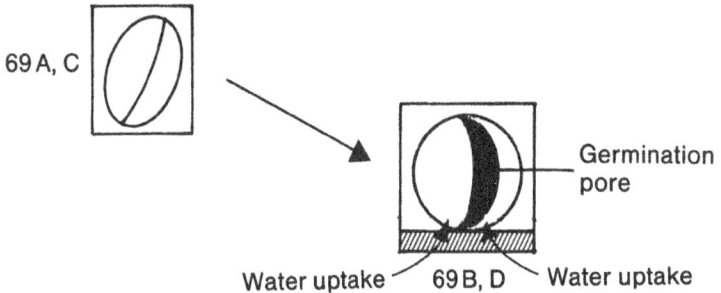

Plate 69A. Pollen grains of *Lilium longiflorum* Thunb. after release from the anther. During the final maturation and opening of the anther, its water content drastically diminishes and the pollen grains become strongly dehydrated, causing them to shrink considerably. x 2,000.

Plate 69B. Pollen grain of *Lilium longiflorum* Thunb. after rehydration. After the pollen has been deposited on the stigma, it starts to take up water from the stigmatic exudate. This leads to strong swelling of the pollen grain, as can be seen from the expansion of the aperture. x 1,200.

Plate 69 C/D. Pollen grains of *Iris florentina* L.
C. shows the pollen grain as it is released from the anther. The aperture is folded together and is visible as a narrow slit. x 950
D. shows the pollen grain after rehydration. It is strongly swollen and the aperture is much wider. x 850

(Plates 69A-D courtesy F Ciampolini, Siena).

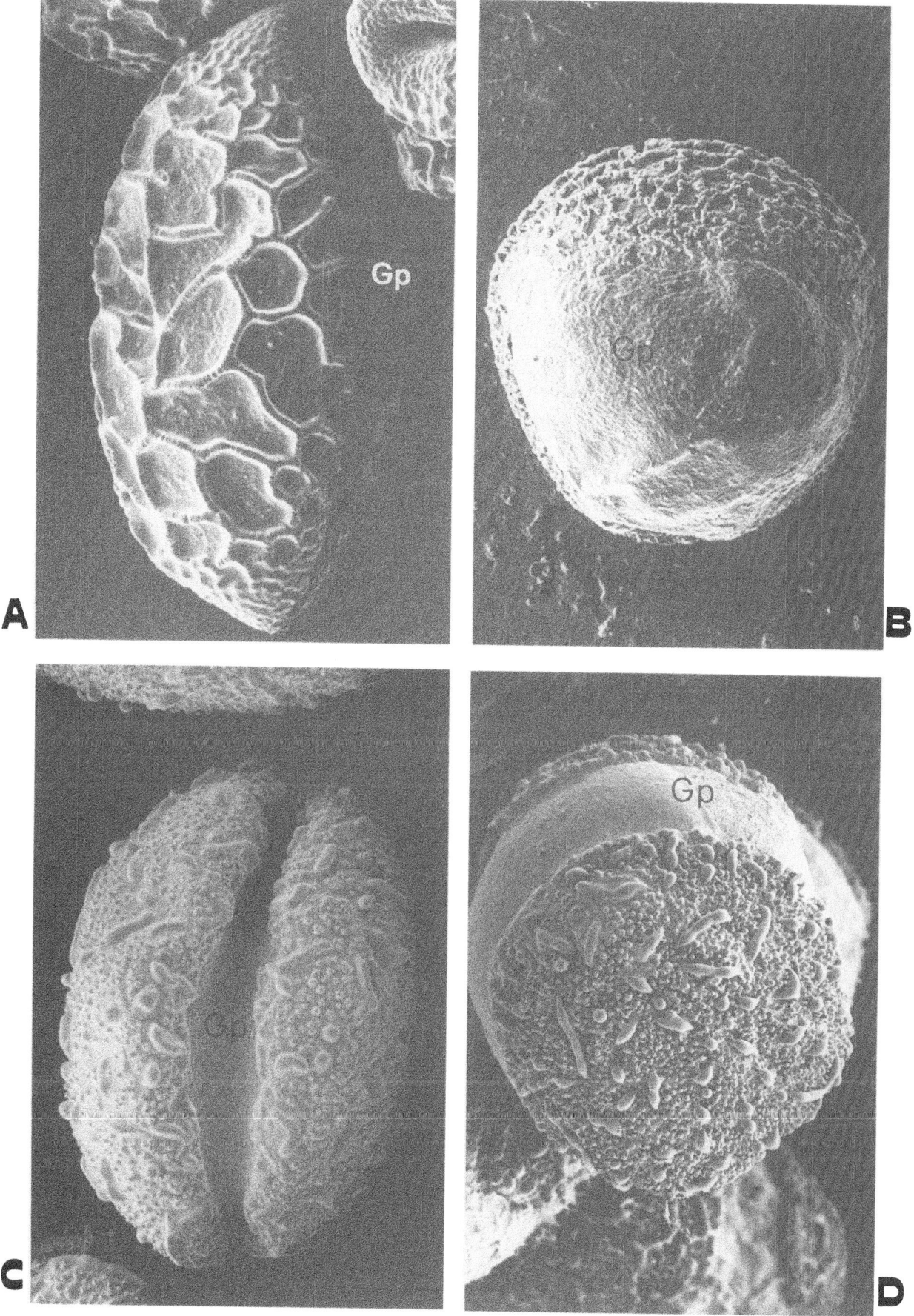

175

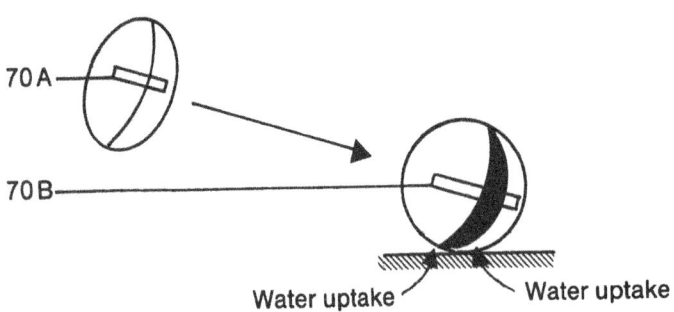

Plate 70A. Ultrastructure of the mature, dehydrated pollen of *Linaria vulgaris* Mill., showing the vegetative nucleus, generative cell and the surrounding cytoplasm. Stacks of rough endoplasmic reticulum are very prominent. x 17,400.

Plate 70B. Ultrastructure of the pollen of *Lycopersicon peruvianum* (L.) Mill. at 1.5 hour after pollination. The pollen has been hydrated and is activated. The rough endoplasmic reticulum, which was originally organized in stacks, has now spread out throughout the entire cytoplasm. x 3,800.

(Plate 70A reproduced by permission from M Cresti et al (1988) Acta Bot Neerl 37: 379-386; Plate 70B courtesy F Ciampolini, Siena).

176

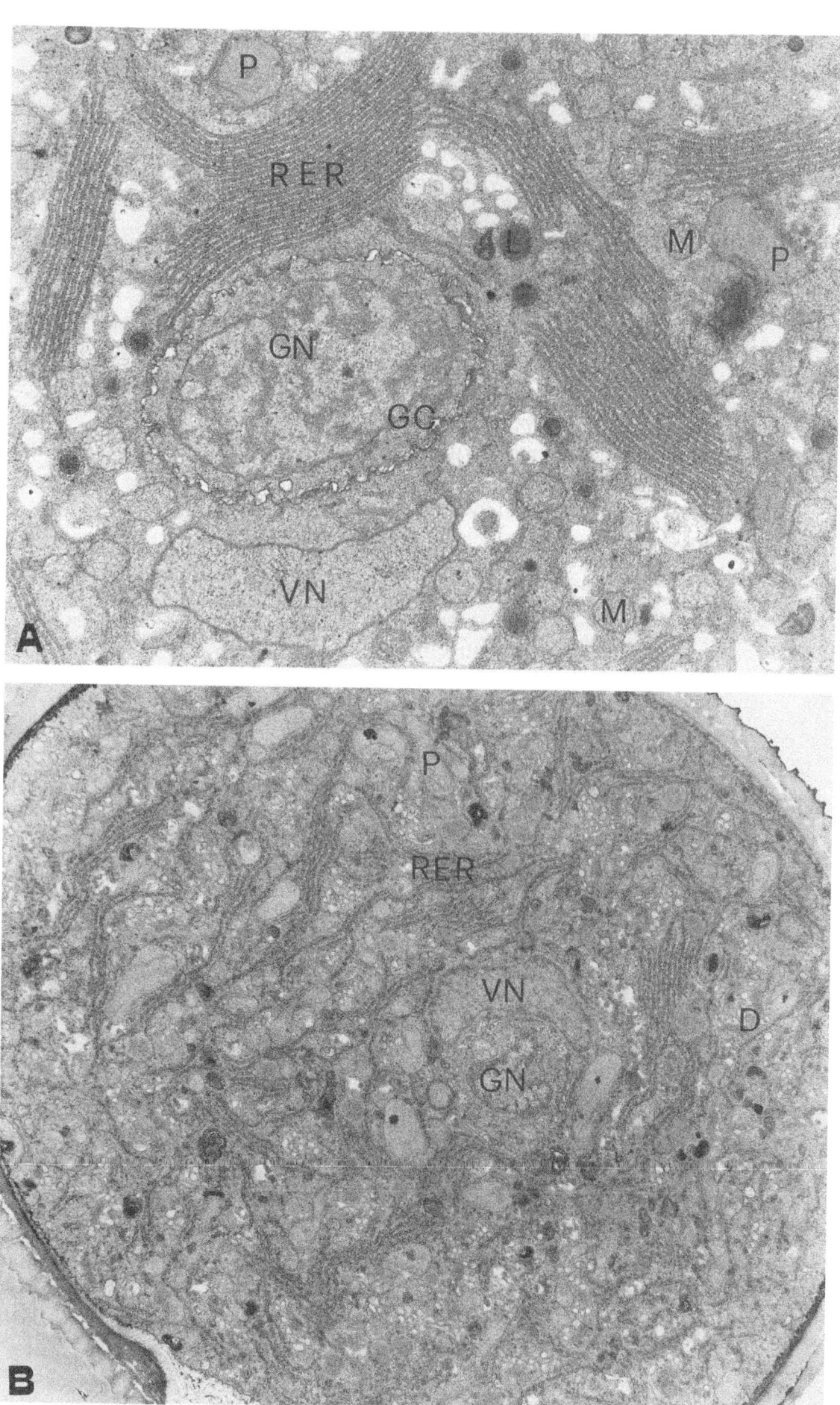

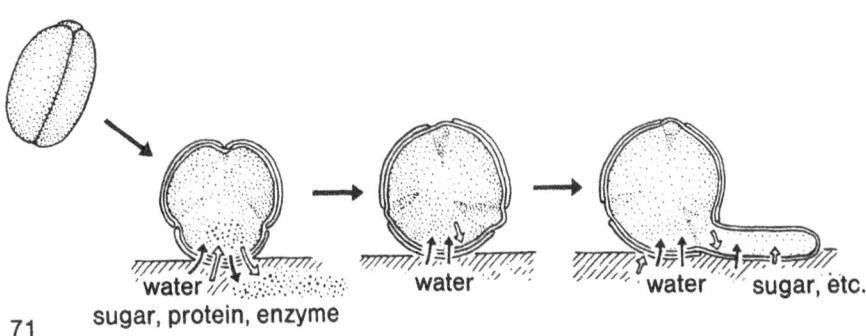

71

Plate 71. Germinating pollen of *Impatiens glandulifera* Royle. The pollen grains of *Impatiens* have four apertures. Some pollen grains have not yet been rehydrated, and their apertures are merely narrow slits (arrows). During the rehydration all four apertures expand (double arrows), a pollen tube is produced by only one of them. x 2,300.

(Plate 71 courtesy A van Aelst, Wageningen).

179

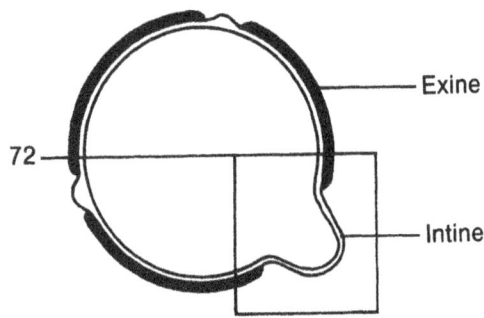

Plate 72. Pollen grain of *Lycopersicon peruvianum* (L.) Mill. germinating on the stigma. The pollen grain is completely rehydrated and activated. Its swollen vegetative cell is protruding through the aperture, but has not yet passed beyond the thickened aperture wall. The cytoplasm contains numerous dictyosomes, which are producing Golgi-vesicles. x 18,000.

(Plate 72 courtesy F Ciampolini, Siena).

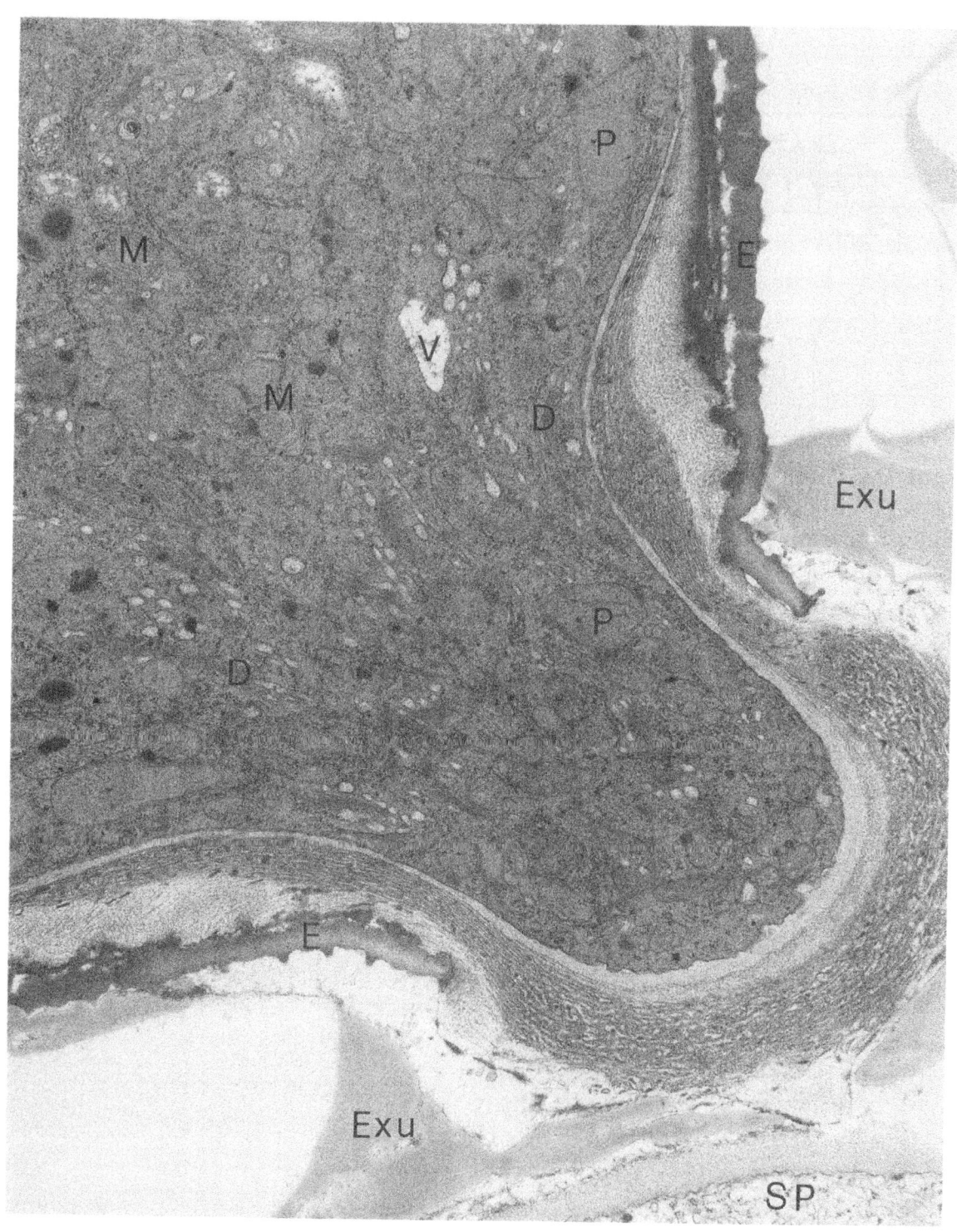

181

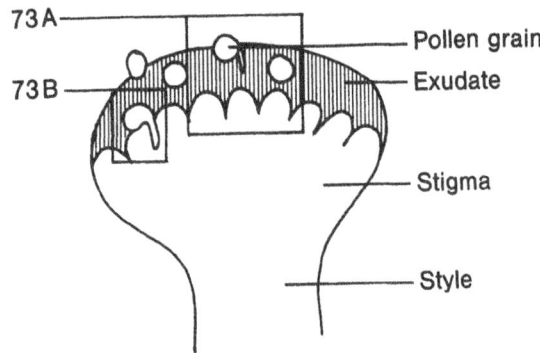

Plate 73A. Pollinated stigma of *Nicotiana tabacum* L.. The material has been critical point dried for scanning electron microscopy. The pollen grains are fully hydrated and some are already germinated, as indicated by the presence of pollen tubes (arrow). x 600.

Plate 73B. Pollen grain of *Lycopersicon peruvianum* (L.) Mill., germinated on the stigma. The vegetative cell has formed a pollen tube which has grown through the wall (arrows) covering the aperture. The vegetative nucleus and the generative cell are still inside the pollen grain. x 6,800.

(Plate 73A reproduced by permission from M Cresti et al (1986) Amer J Bot 73: 1713-1722; Plate 73B original M Cresti).

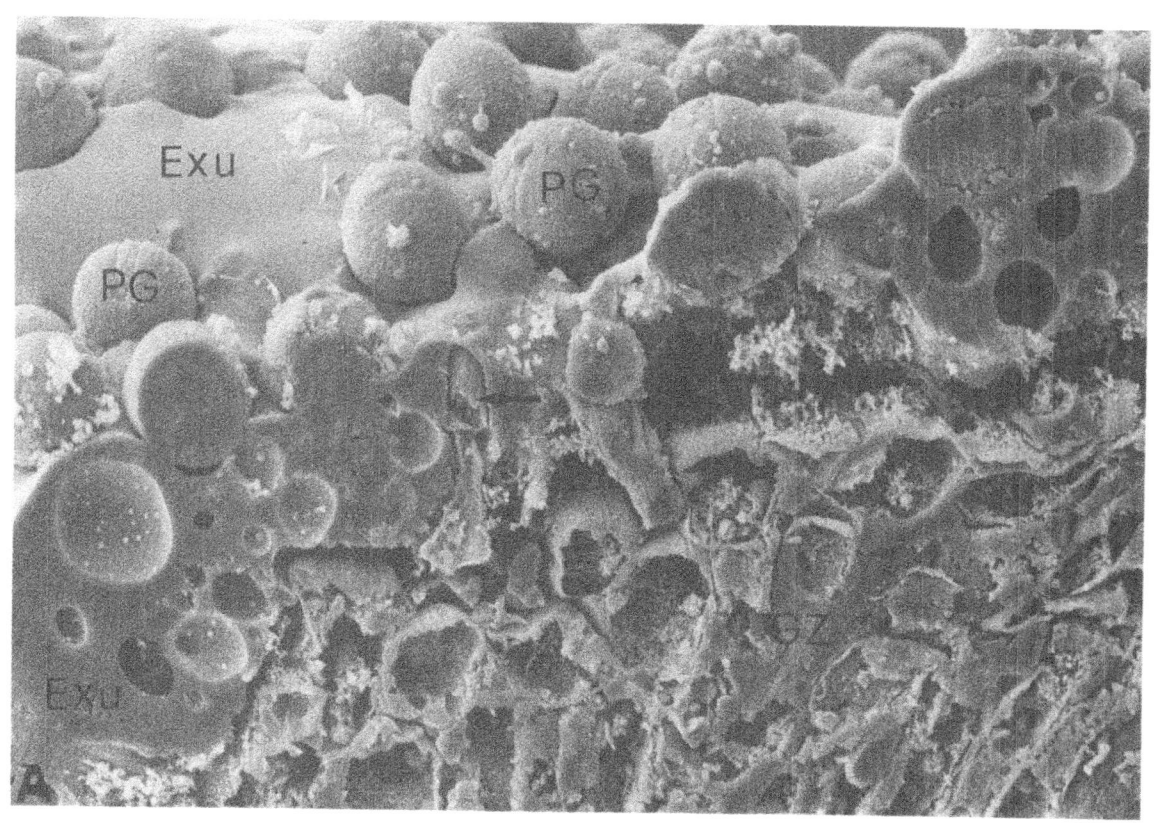

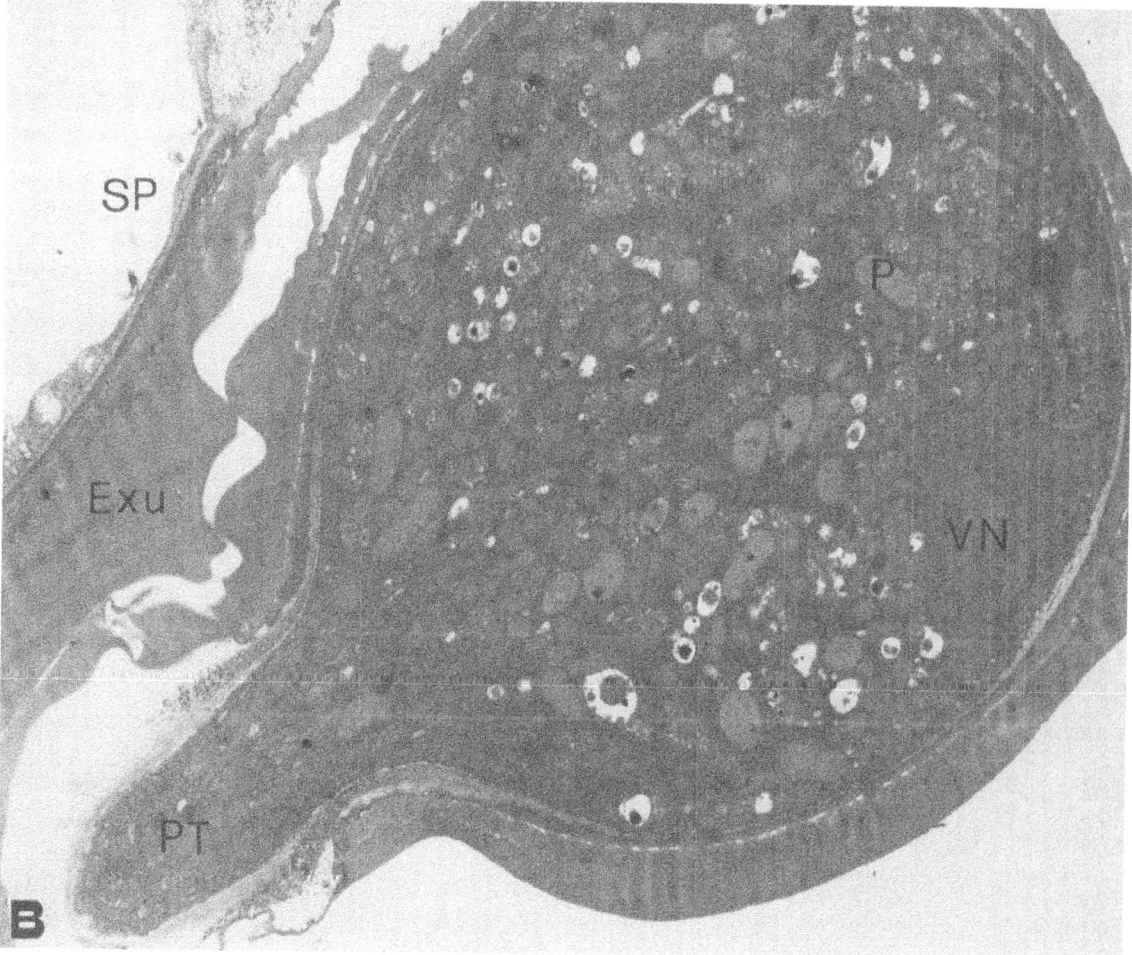

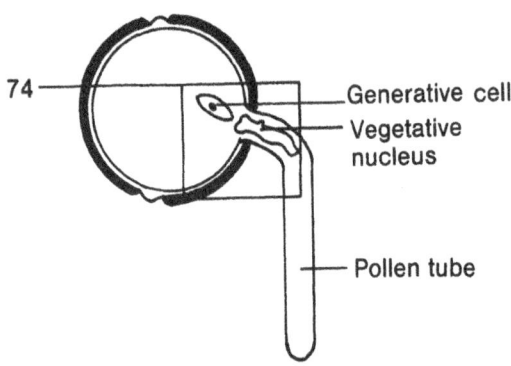

Plate 74. Enlarged portion of germinated pollen grain and pollen tube of *Lycopersicon peruvianum* (L.) Mill. The vegetative nucleus and generative cell are about to enter the pollen tube. The cytoplasm of the vegetative cell contains large numbers of Golgi-vesicles. In the portion of the vegetative cell which is still inside the pollen grain, vacuoles are formed. x 9,300.

(Plate 74 courtesy F Ciampolini, Siena).

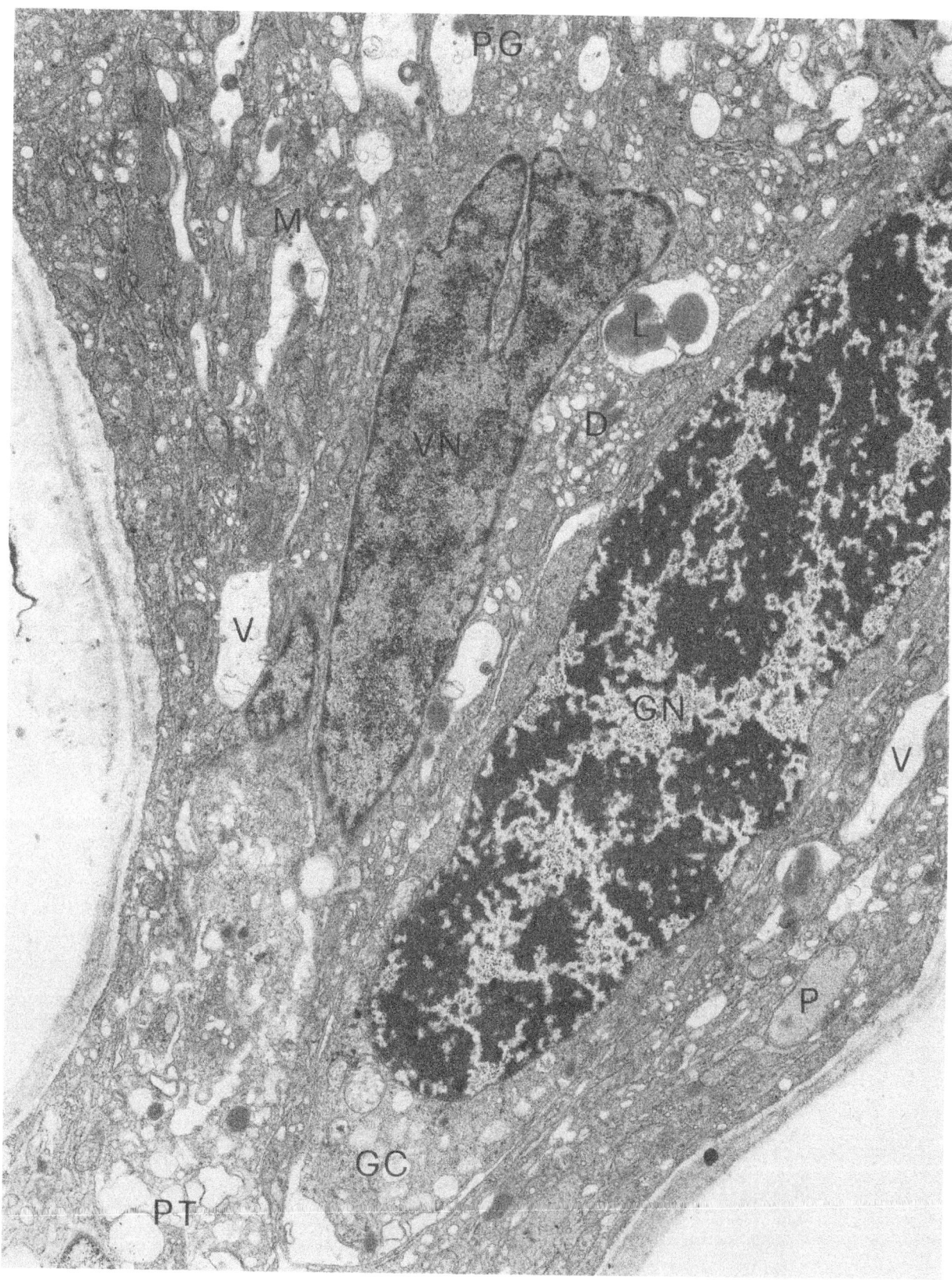

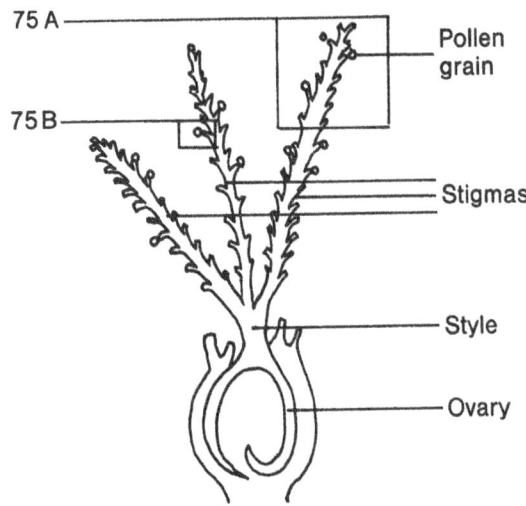

75 A

75 B

Pollen grain

Stigmas

Style

Ovary

Plate 75A. Pollinated stigma of *Spinacia oleracea* L.. The stigma of *Spinacia* is of the dry type, and the pollen grains adhere to the stigma papillae. Some grains are already hydrated and swollen (asterisks), and have formed a pollen tube (arrow). x 1,100.

Plate 75B. Section through a pollinated stigma of *Spinacia oleracea* L.. After hydration and activation the pollen grain forms a pollen tube, which grows along the papillar surface towards the style. The pollen tube penetrates the cuticle of the papilla (arrow) and continues its growth through the wall. x 2,300.

(Plates 75A, B reproduced by permission from H Wilms (1980) Acta Bot Neerl 29: 33-47).

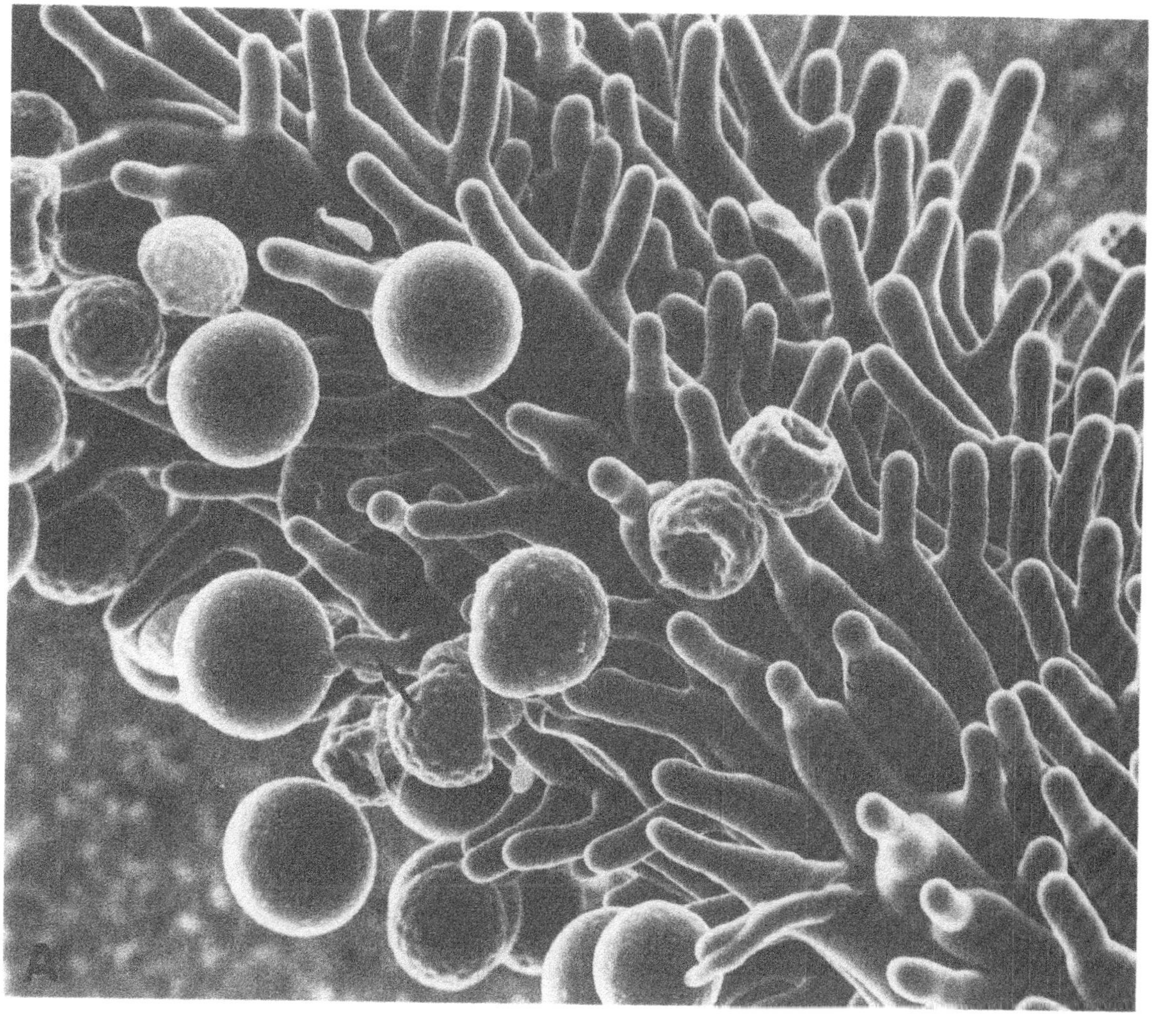

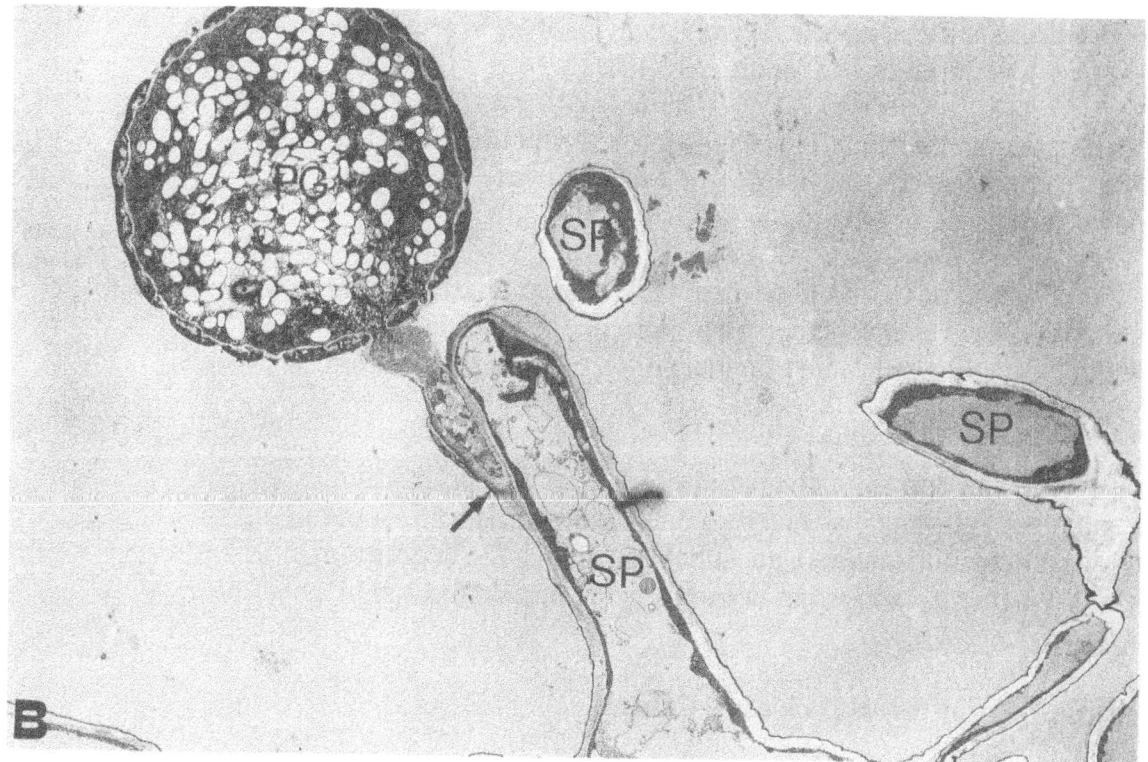

Plates 76A-H. Sporophytic self-incompatibility in *Brassica oleracea* L..

A-C. Development of compatible pollen on the stigmatic papilla viewed in vivo. Frames were recorded at intervals after pollination; A: 5 mins; adhesion, B: 40 mins; hydration, and C: 110 mins; tube entry. x 1,000.

D. Development of compatible and incompatible pollen grains on the same stigmatic papilla 45 mins after pollination. The incompatible grain remains dehydrated, whilst the compatible pollen grain has hydrated and begun to produce a tube. x 800.

E. Pollen-stigma interface 60 mins after a compatible pollination. The pollen coat (asterisks), previously homogenous, has been converted into an electron opaque matrix which serves to attach the grain firmly to the stigmatic papilla. x 16,400.

F. Pollen stigma interface 24 h following a self-pollination. The pollen coating (asterisks) is incompletely converted and has become detached from the stigmatic surface. x 14,400.

G. Pollen grain on a stigmatic papilla 6 h after a compatible pollination. The tube, together with most of the cytoplasm, has entered the papilla wall. Note a further pollen tube in cross section. x 2,900.

H. Pollen grains and stigmatic papilla 24 h after self-pollination: the tube, now necrotic, has failed to enter the papilla wall. Some cytoplasmic reaction (arrows), presumably involving the deposition of the 1-3 ß glucan, callose, has occurred in the papilla. x 6,300.

(Plate 76A-H courtesy H Dickinson, Oxford).

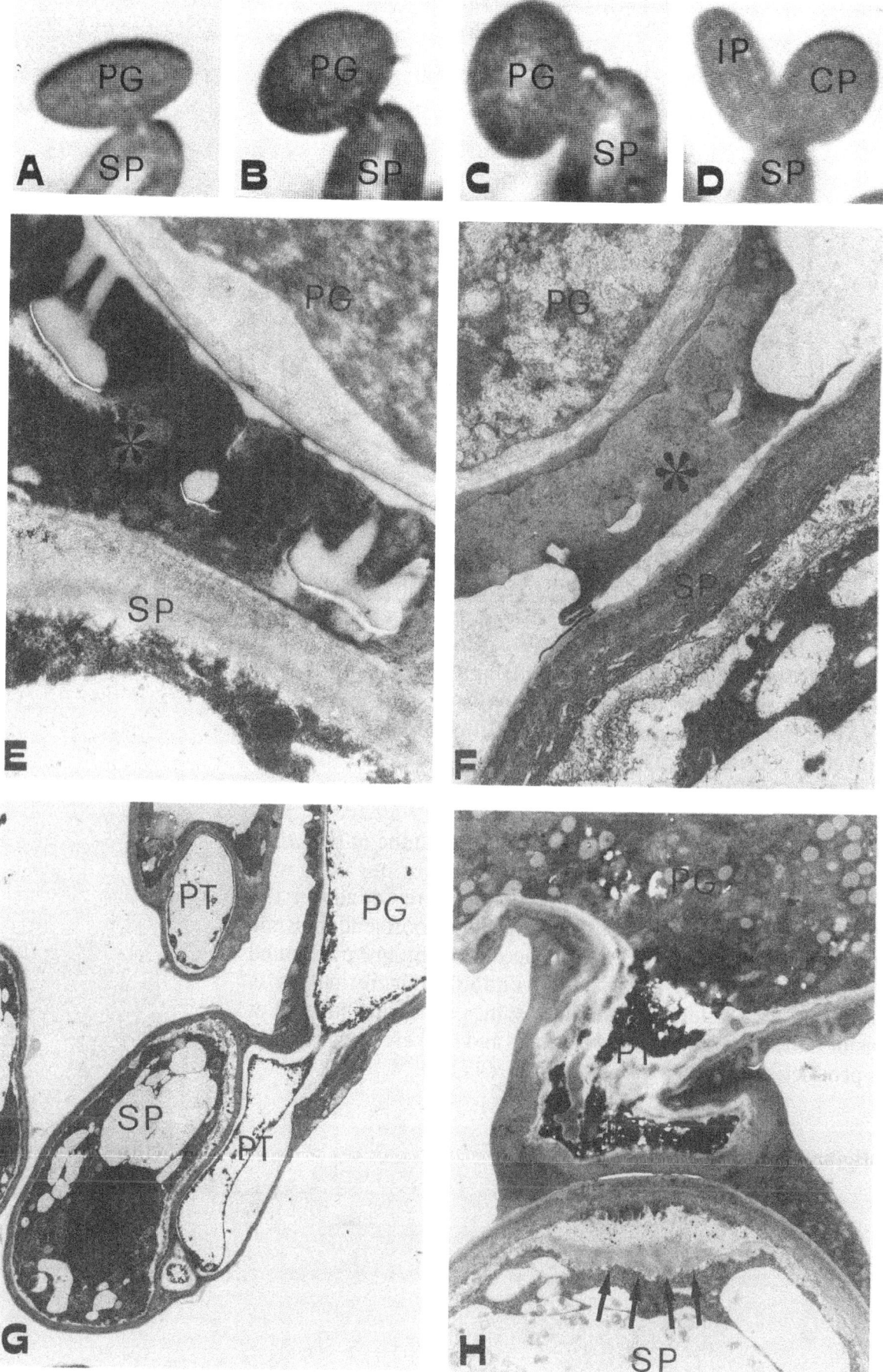

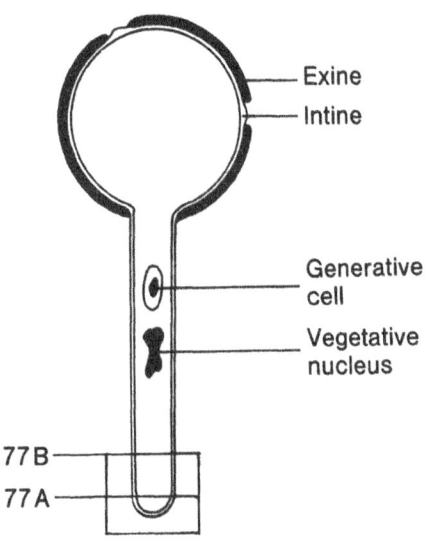

Exine
Intine

Generative cell

Vegetative nucleus

77B
77A

Plate 77A. Longitudinal section through the apical region of a pollen tube of *Impatiens walleriana* Hook f. Usually, pollen tubes show a distinct zonal organization, with an apical or tip region, a subapical zone, a nuclear zone and a vacuolated region, which together form the living part of the pollen tube. The living part of the pollen tube is separated from the non-living part by a callose plug. The tip region primarily contains Golgi-vesicles, which are involved in cell wall formation at the tube apex. x 35,000.

Plate 77B. Longitudinal section through a pollen tube of *Impatiens walleriana* Hook f., near the tip region. In this region, along with numerous Golgi-vesicles, also mitochondria are found. In the *Impatiens* pollen tubes an extensive system of smooth endoplasmic reticulum is present, of which large accumulation are positioned near the tip regions. This smooth endoplasmic reticulum is continuons with sistems of the dictyosomes which are located at considerable distance from the tube tip, and it likely is involved in the production of Golgi-vesicles on site. x 45,000.

(Original J Van Went).

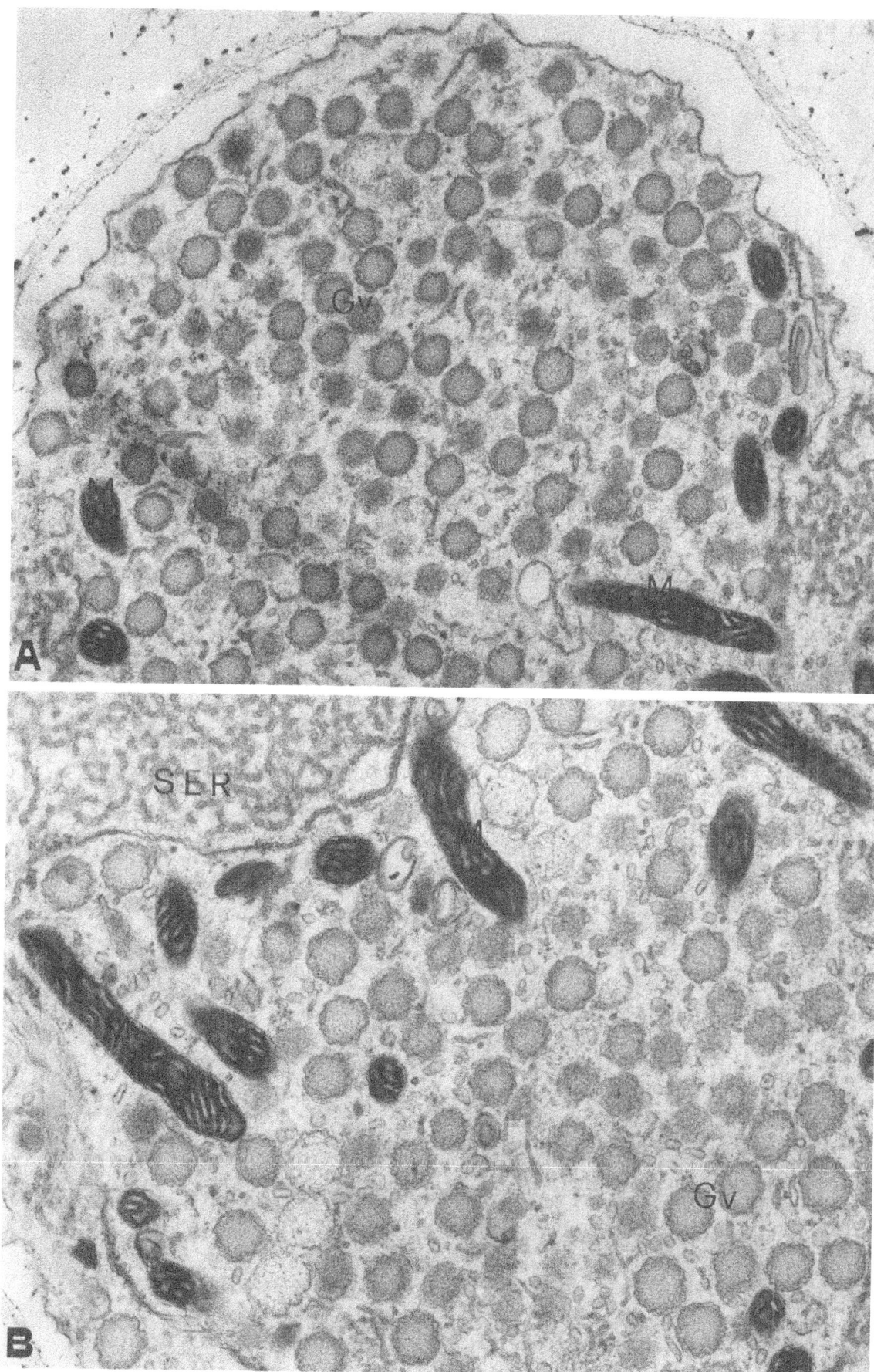

191

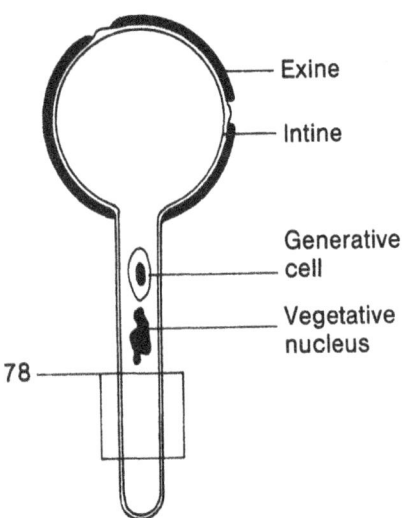

Exine

Intine

Generative cell

Vegetative nucleus

78

Plate 78. Transverse section through the subapical region of a pollen tube of *Nicotiana alata* Link & Otto. Like the tip region, this portion of the pollen tube also is very rich in cytoplasm. Very prominent in this region are the numerous dictyosomes and mitochondria, as well as many Golgi-vesicles. The many dictyosomes are involved in the production of Golgi-vesicles (arrow), through which polysaccharidic cell wall precursors are transported to the tip region. x 135,000.

(Plate 78 courtesy F Ciampolini).

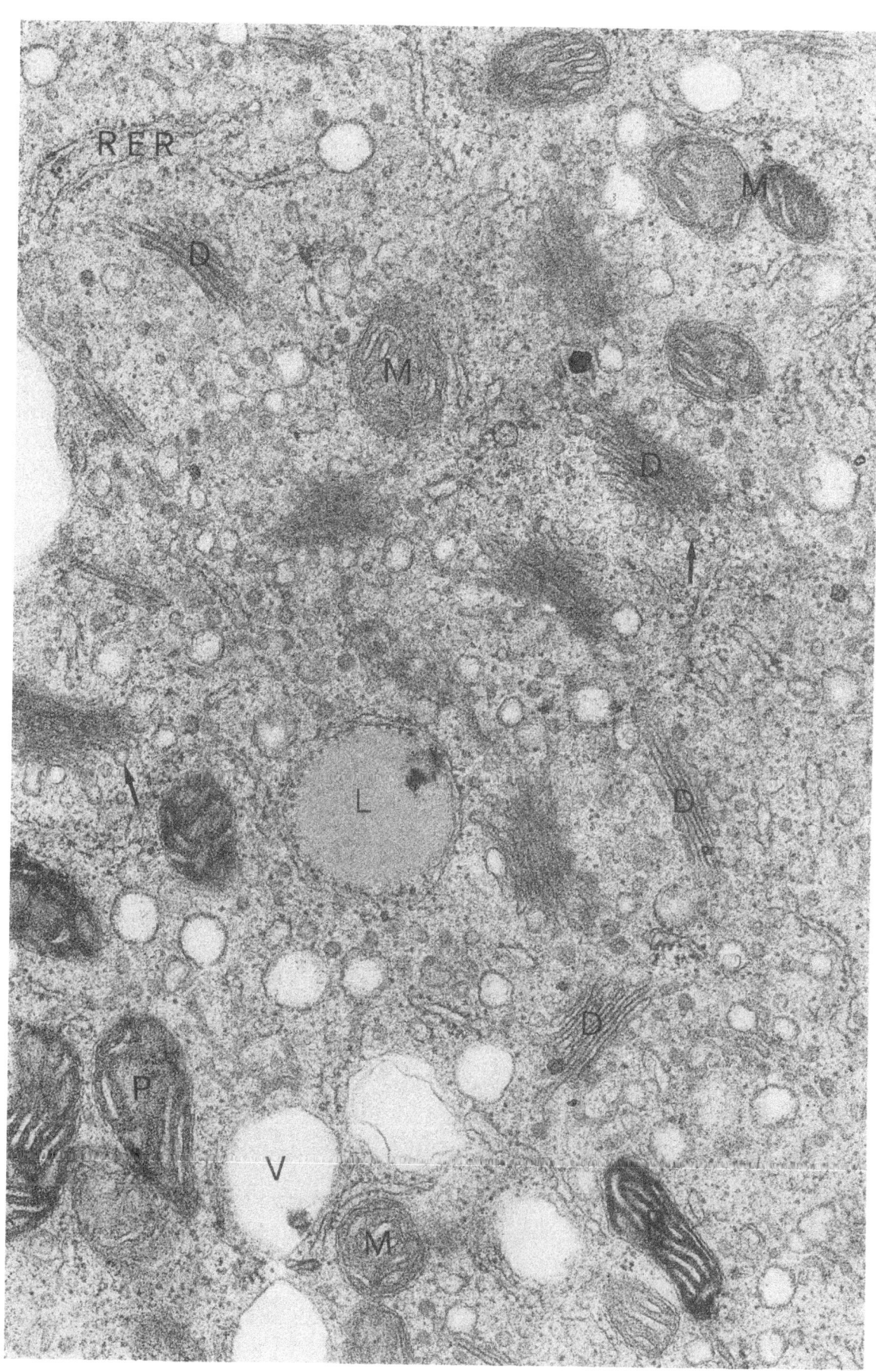

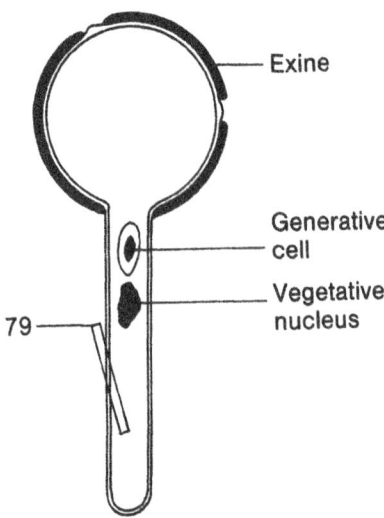

Plate 79. Longitudunal section of freeze-fixed and freeze-substituted pollen tube of *Nicotiana alata* Link & Otto, showing the orientation and position of the cortical microtubules of the vegetative cell. The microtubules run parallel to the plasma membrane and the longitudinal axis of the tube. Fine filaments (arrows) running parallel to microtubules are present. x 80,000.

(Plate 79 courtesy P Hepler, Amherst).

194

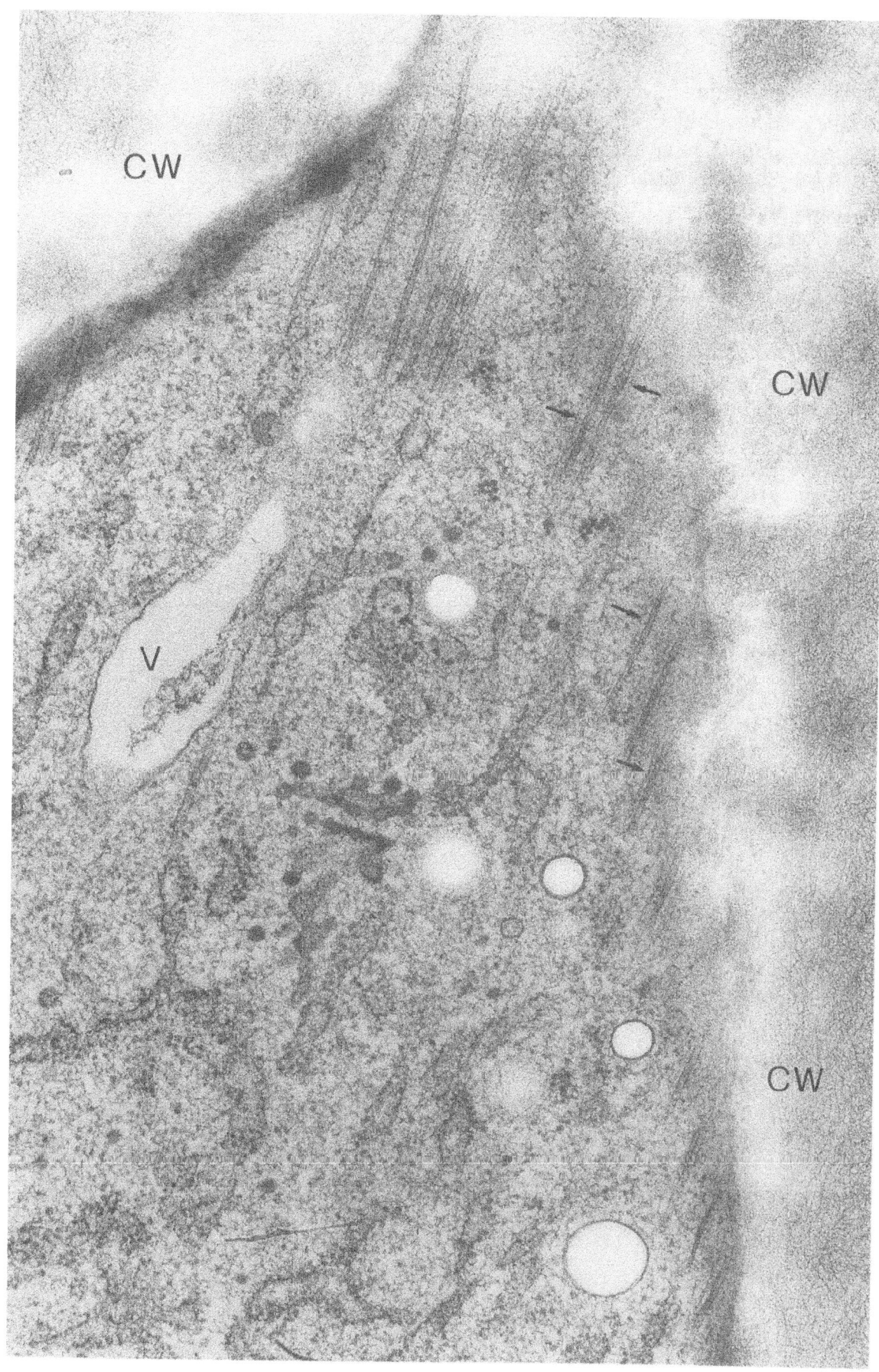

195

Plate 80. Visualization of cytoskeletal elements in pollen tubes of *Nicotiana tabacum* L. by UV fluorescence microscopy.
A. Microfilaments near the surface of the generative cell. Rhodamine-Phalloidine staining. x 470.
B. Microtubules come out from the grain and elongate with cortical distribution through the pollen tube. Immunostaining: as primary antibody a monoclonal to the mammalian tubulin subunit and a FluoresceinIsoThioCyanate labelled rabbit anti mouse as secondary antibody. x 420.
C, D, E. Microfilaments at different focal plane inside the same pollen tube. Rhodamine-Phalloidine staining. x 330.

(Plates 80A-E courtesy A Tiezzi, Siena).

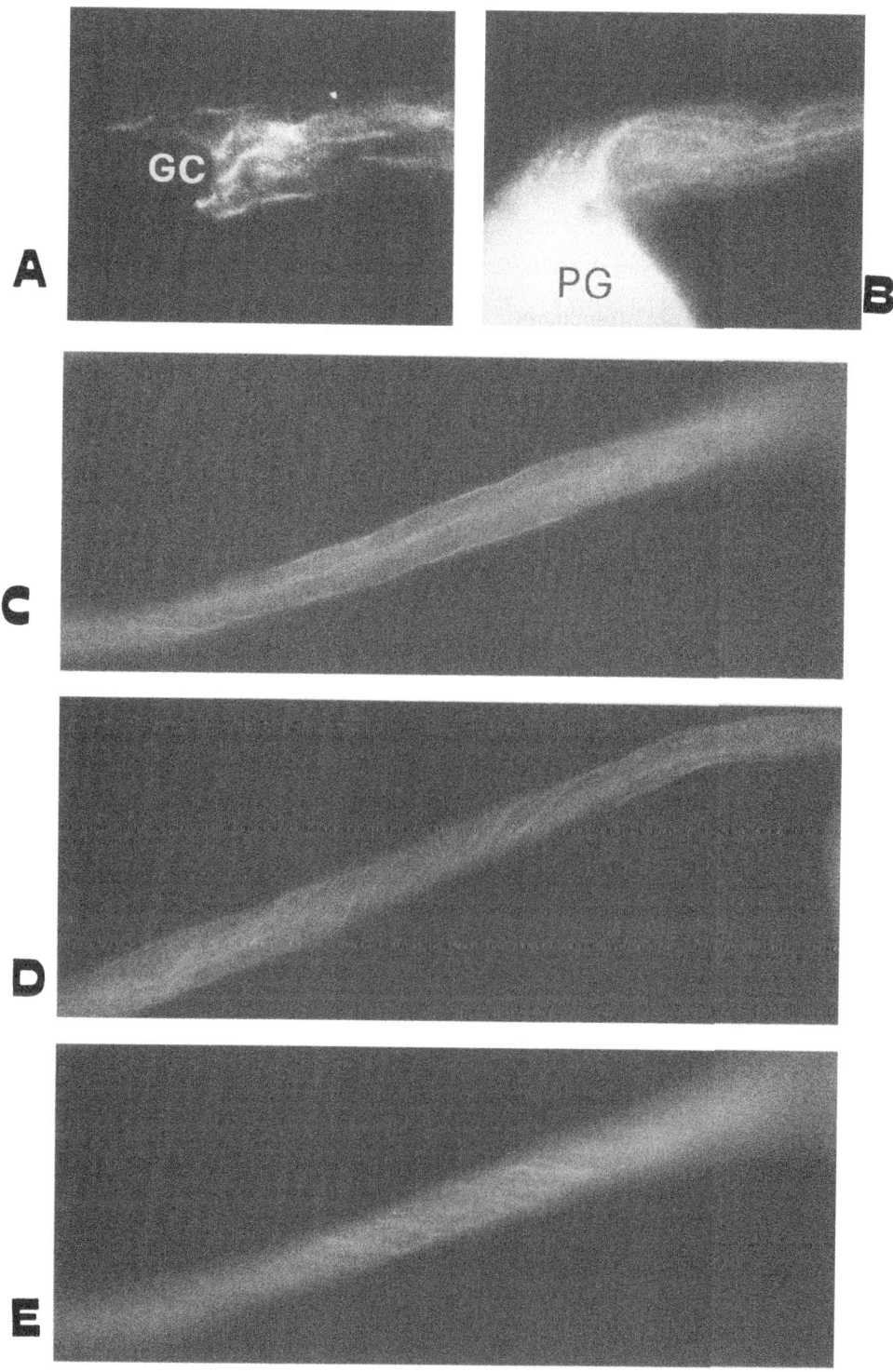

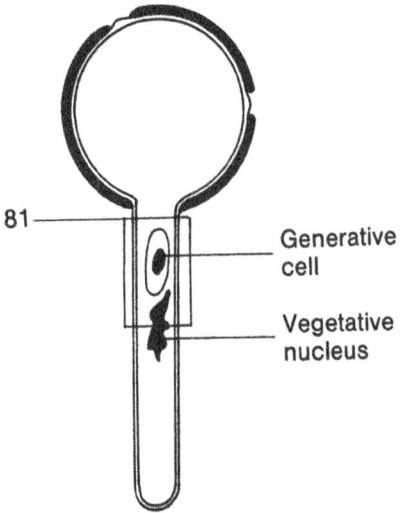

Plate 81. Portion of a pollen tube of *Nicotiana alata* Link & Otto, showing the vegetative nucleus and the generative cell. The material has been freeze-fixed and freeze-substituted. The pollen tube has a bilayered wall in this tube region: an external pectocellulosic wall layer, and an inner callosic layer (arrows). The vegetative nucleus is highly irregular in outline, resulting in a number of section profiles. The generative cell remains positioned close to the vegetative nucleus. The arrowheads point bundles of microtubules in the generative cell. x 14,000.

(Plate 81 courtesy P Hepler, Amherst).

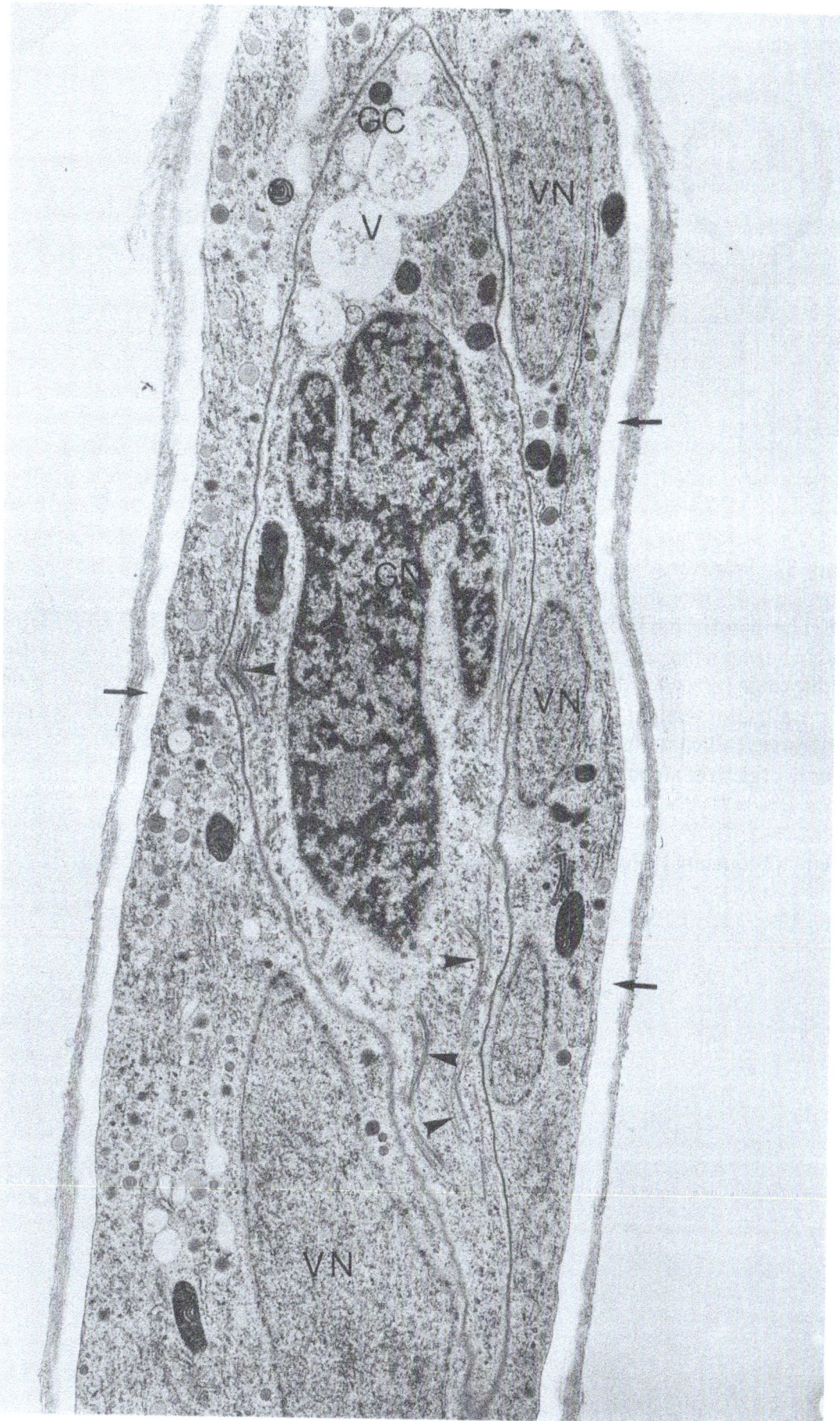

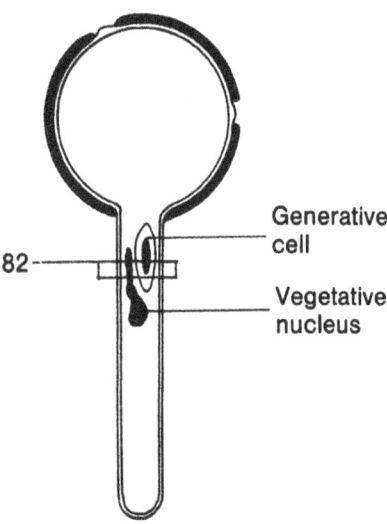

Plate 82. Transverse section through a pollen tube of *Nicotiana alata* Link & Otto, showing the vegetative nucleus and generative cell. The material has been freeze-fixed and freeze-substituted. The wall in between the vegetative and generative cell has a very regular in thickness (arrowheads). The two plasma membranes, bordering this wall, are straight and closely appressed to the wall. The generative cell contains a large number of microtubules (arrows) mainly organized in bundles. x 39,000.

(Plate 82 courtesy P Hepler, Amherst).

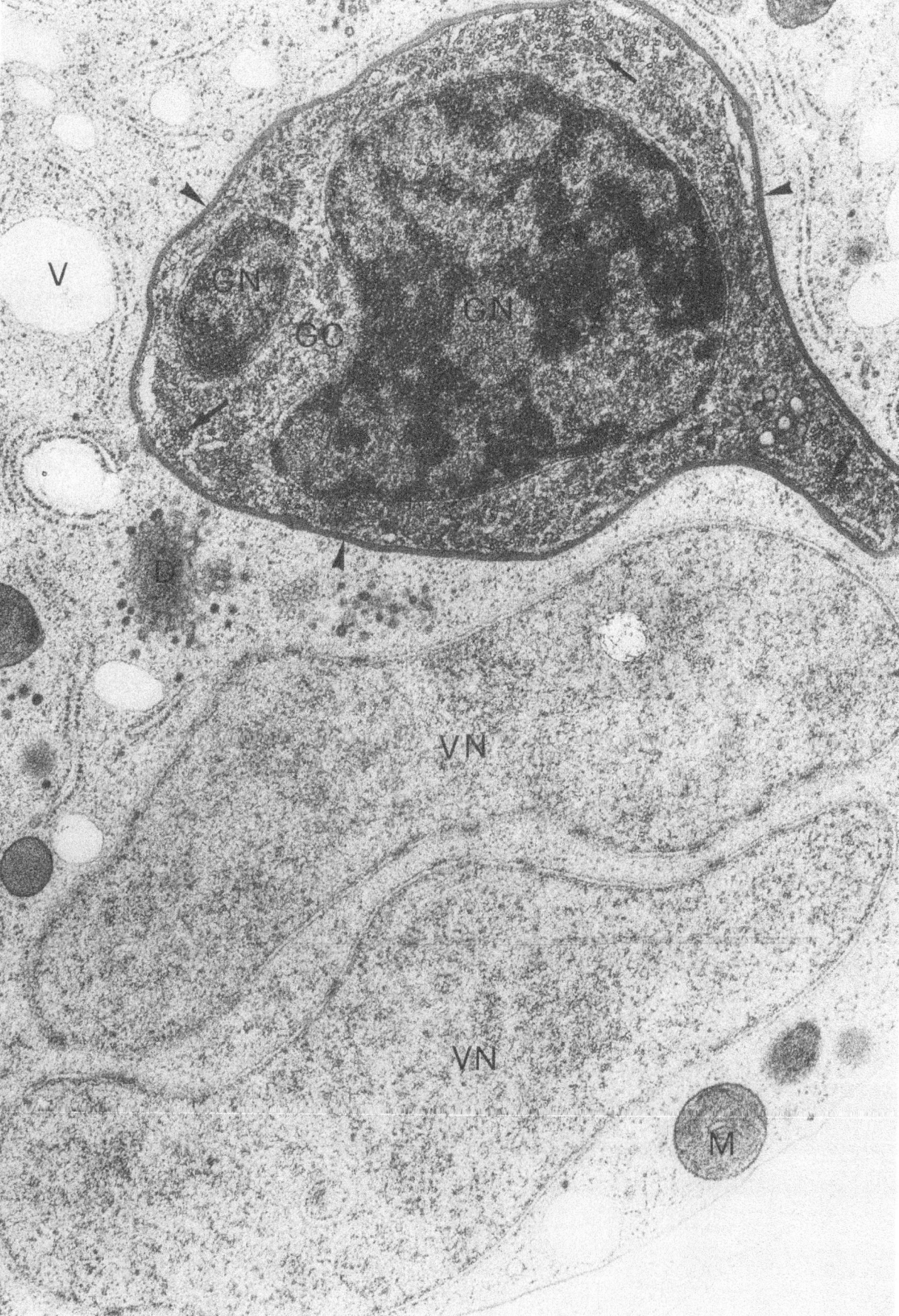

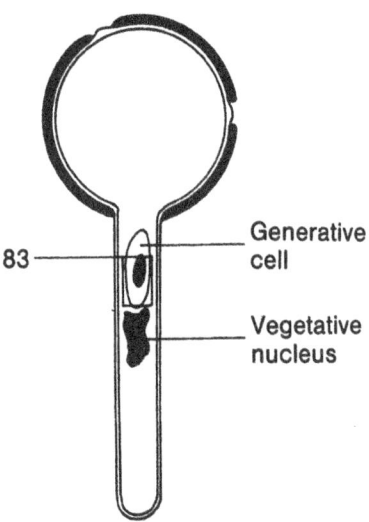

83 —

Generative
cell

Vegetative
nucleus

Plate 83. Longitudinal section through the generative cell of *Nicotiana alata* Link & Otto pollen tube, prepared by freeze-fixation and freeze-substitution. The arrowheads point to the wall between the generative and the vegetative cell. x 68,000.

(Plate 83 courtesy P Hepler, Amherst).

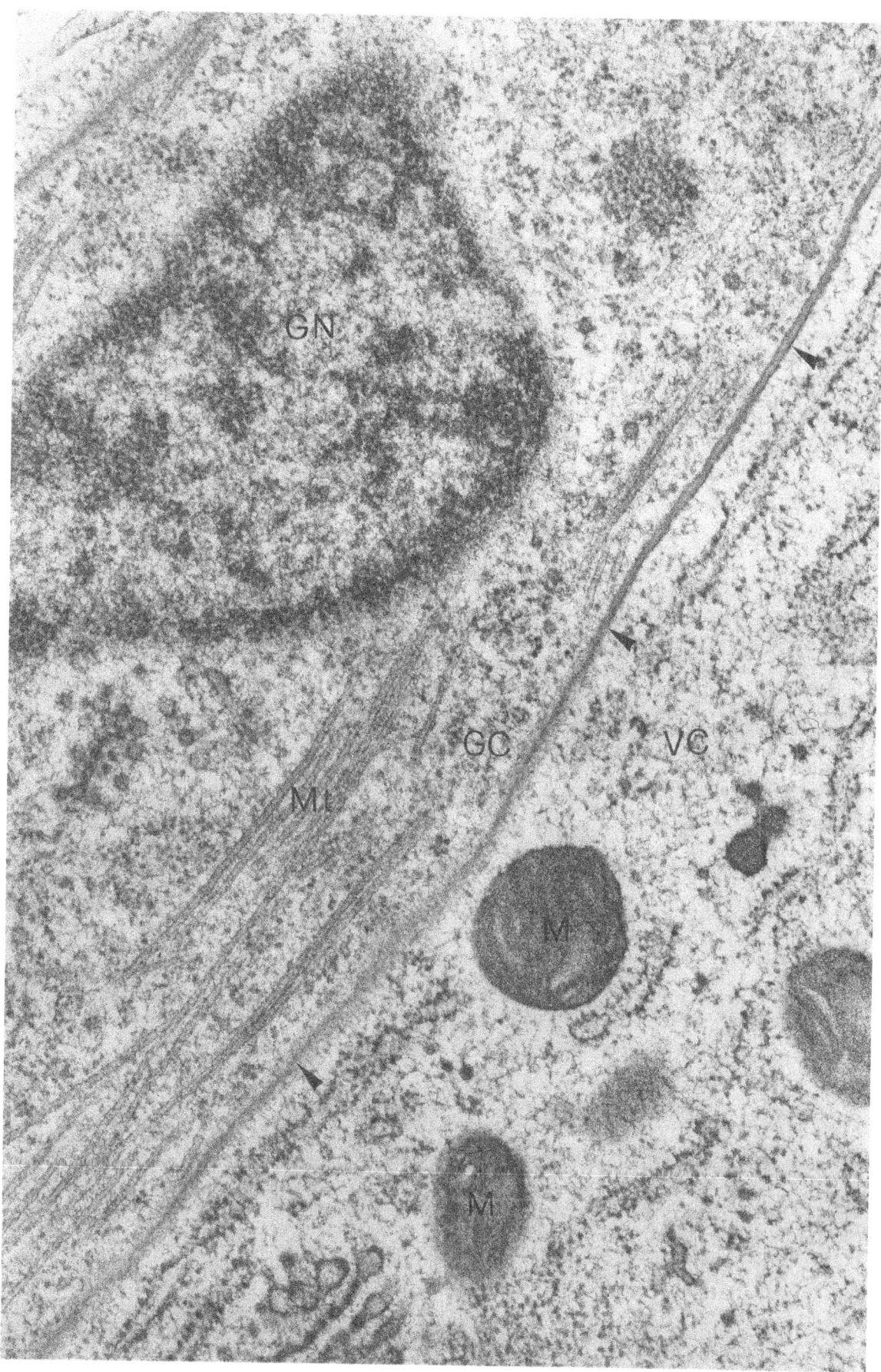

203

Plate 84. Immunolocalization of myosin in pollen grains and pollen tubes.

A. Vegetative nucleus of a pollen grain from *Hyacinthus orientalis* L., seen with Differential Contrast Microscopy (DIC). x 1,800. **B.** Same nucleus as A, showing labeling of myosin at the nuclear envelope. x 1,800. **C.** Immunofluorescence localization of myosin on the vegetative nucleus of *Helleborus foetidus* L. x 2,500. **D.** Myosin localization on the vegetative nucleus of an activated pollen grain from *Hyacinthus orietnalis* L. x 1,800. **E.** Myosin localization on the elongated vegetative nucleus from a pollen tube of *Hyacinthus orientalis* L. x 2,000. **F.** Generative cell of a pollen grain from *Helleborus foetidus* L., as seen with DIC. x 2,500. **G.** Same cell as **F.**, showing surface labeling of myosin. x 2,500. **H.** Immunolocalization of myosin in the tip region of a pollen tube of *Nicotiana alata* Link & Otto. x 3,000. **I.** Distribution of myosin in the tube region near the generative cell (arrow). x 3,000. **K.** Localization of myosin associated with the surfaces of the vegetative nucleus and generative cell of *Nicotiana alata* Link & Otto pollen tube. x 3,000. **L.** Same portion of the pollen tube as shown in **K.**, but stained for nuclei with DAPI. x 3,000.

(Plates 84A-G reproduced by permission from Heslop-Harrison J, Heslop-Harrison Y (1989) J Cell Sci 94: 319-325; Plates 84 H-L reproduced by permission from X Tang et al (1989) J Cell Sci 92: 569-574).

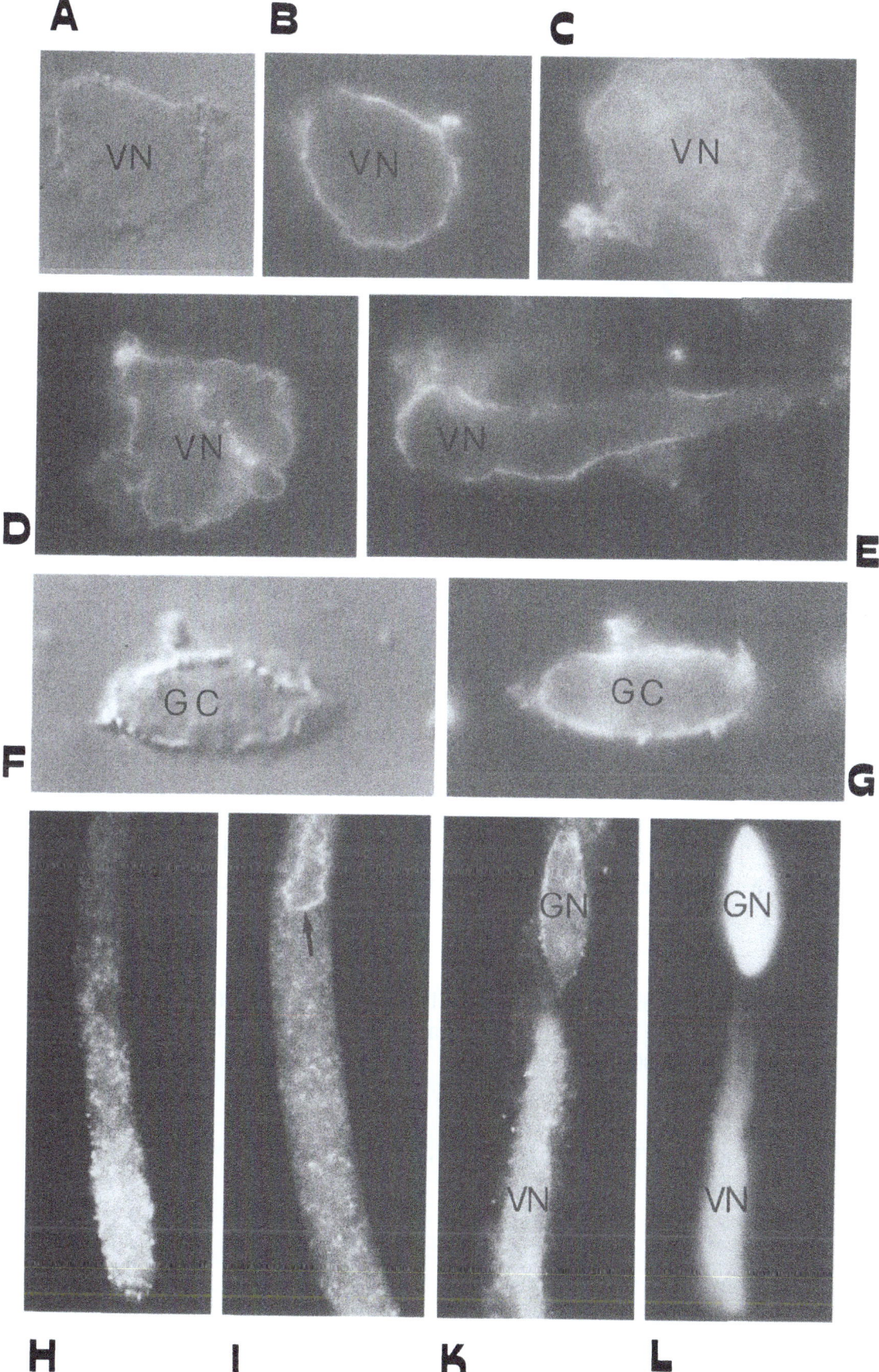

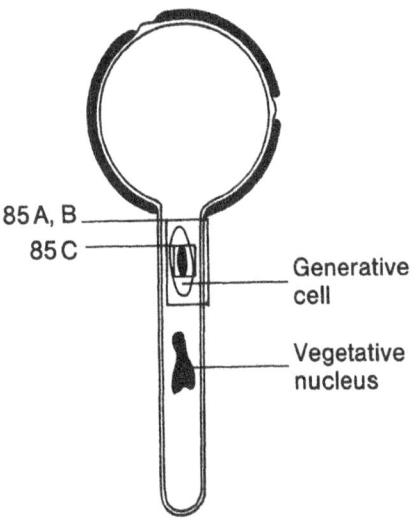

85 A, B
85 C

Generative
cell

Vegetative
nucleus

Plate 85A. Generative cell in the pollen tube of *Nicotiana tabacum* L., showing the cytoskeleton composed of microtubules, visualised by UV-immunolabelling. The microtubules are not randomly distributed, but are arranged in clusters, which form a basket-like structure. x 510.

Plate 85B. The generative cell in the *Nicotiana tabacum* L. pollen tube has a tail-like extension, which also contains microtubules (arrow). x 450.

Plate 85C. Portion of the generative cell in pollen tube of *Aloë ciliaris* Haw., showing the position and arrangement of the cortical microtubules. x 128,000.

(Plates 85A, B courtesy A Tiezzi,Siena; Plate 85C courtesy F Ciampolini, Siena).

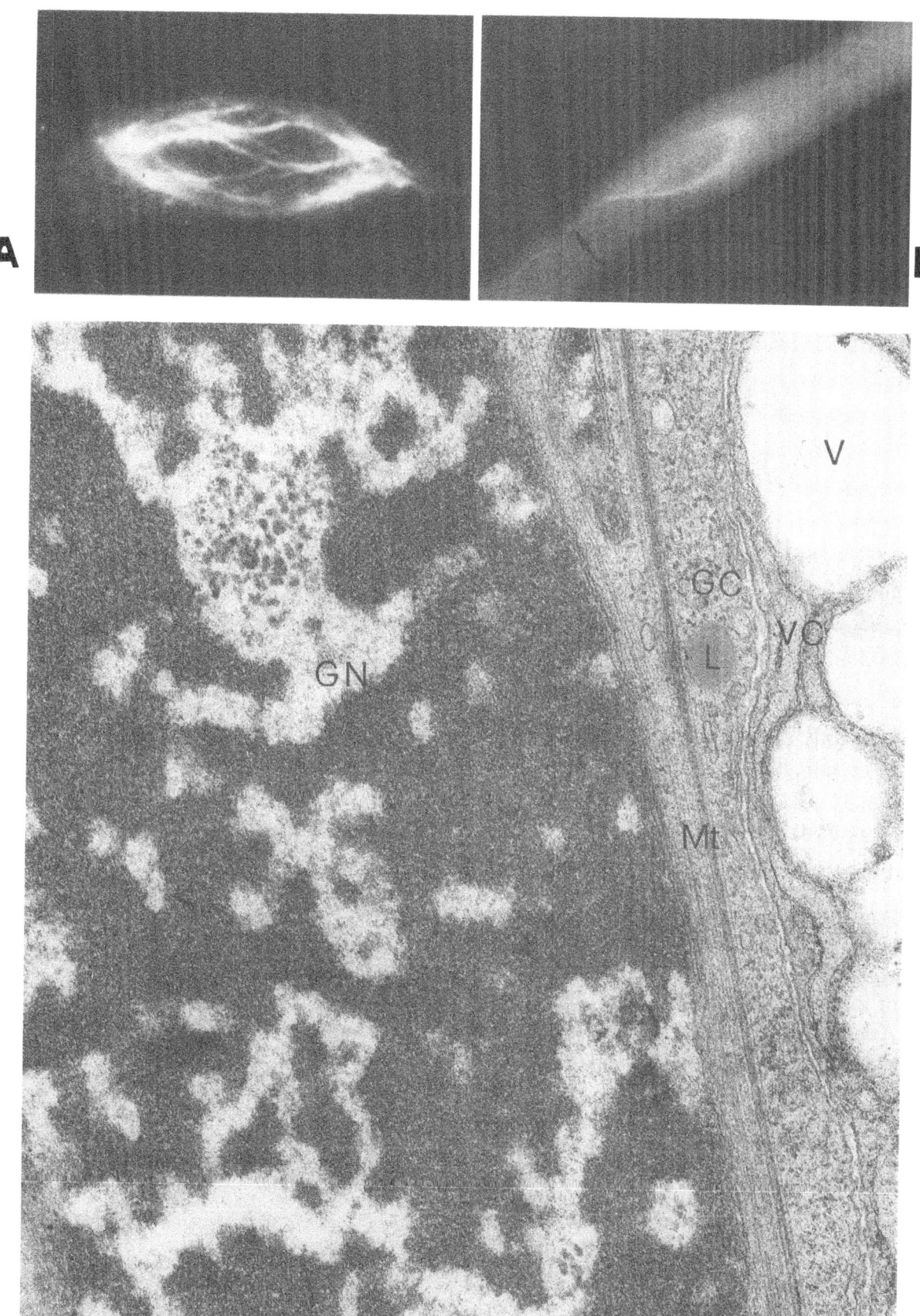

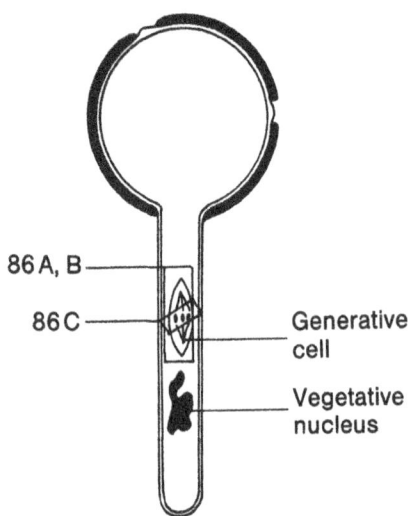

86 A, B

86 C — Generative cell

Vegetative nucleus

Plate 86A. Generative cell in the pollen tube of *Nicotiana tabacum* L. at metaphase of mitosis. The microtubules, visualized by UV-immunolabeling, are arranged in bundles of the spindle-figure. x 500.

Plate 86B. Generative cell in the pollen tube of *Nicotiana tabacum* L. at telophase of mitosis. Only some bundles of microtubules remain present in between the two newly formed nuclei, where cytokinesis will take place (arrow). x 300.

Plate 86C. Generative cell in the pollen tube of *Nicotiana tabacum* L. at metaphase of mitosis. The chromosomes are positioned at the equatorial plane of the cell. In between the chromosomes bundles of microtubules (arrows), forming the spindle figure can be seen. Arrowheads point to the plasmamembranes in between the vegetative cell and the generative cell. x 74,000.

(Plates 86A, B courtesy A Tiezzi, Siena; Plate 86C courtesy F Ciampolini, Siena).

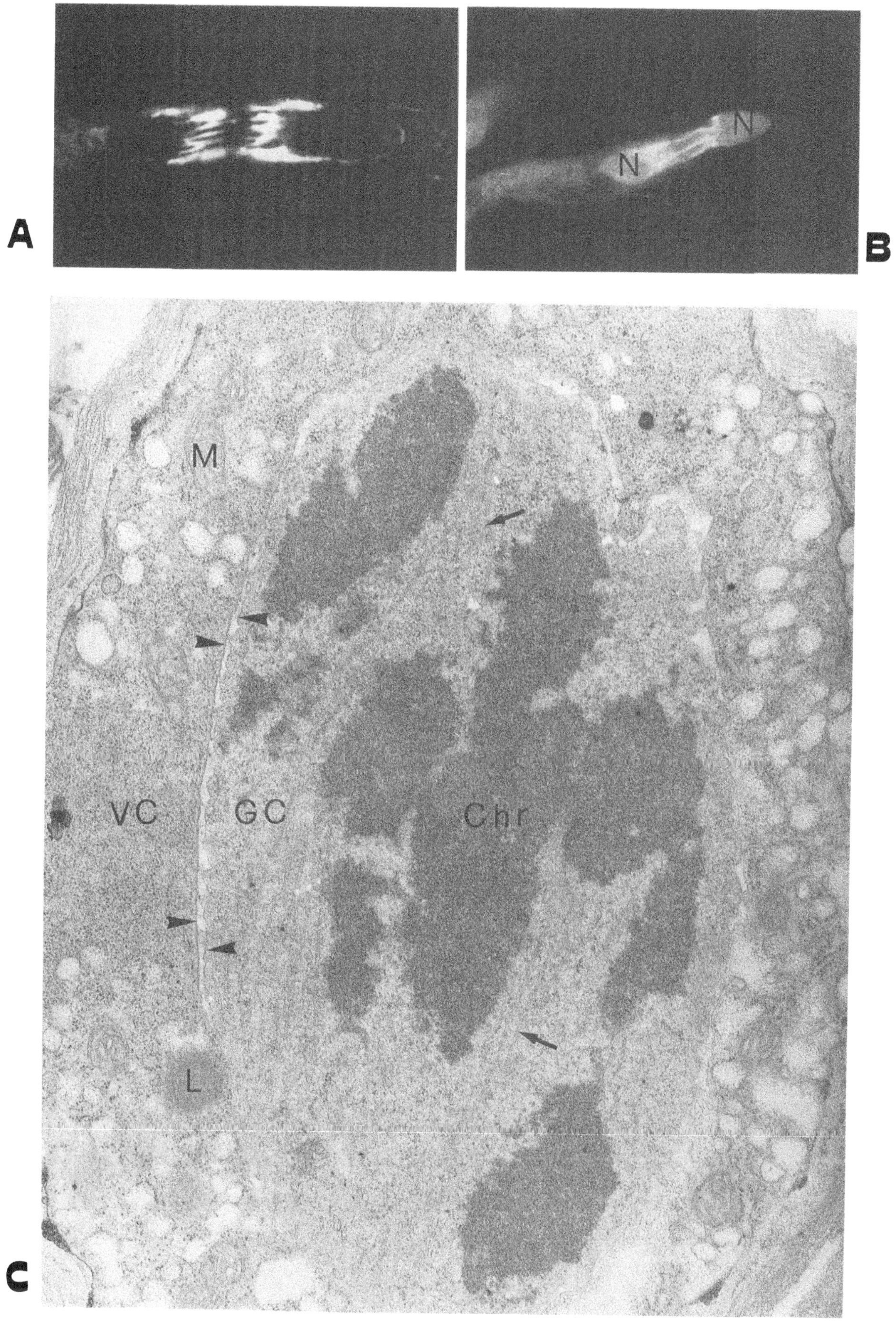

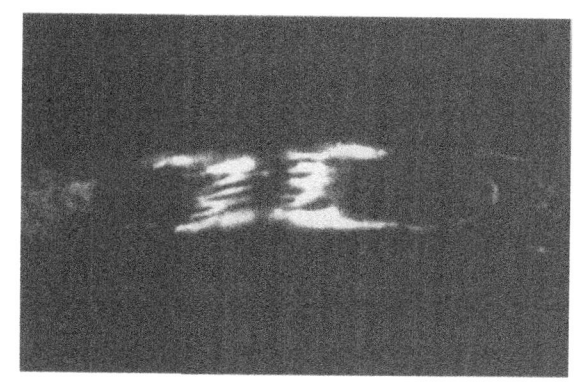

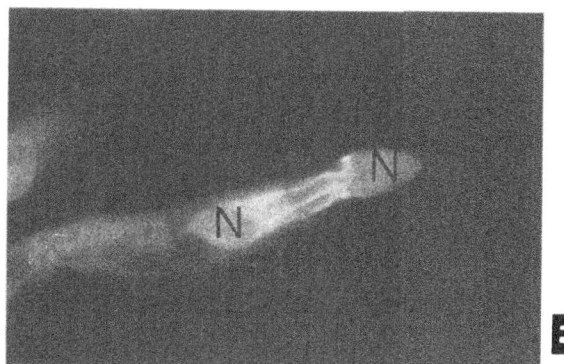

A

B

C

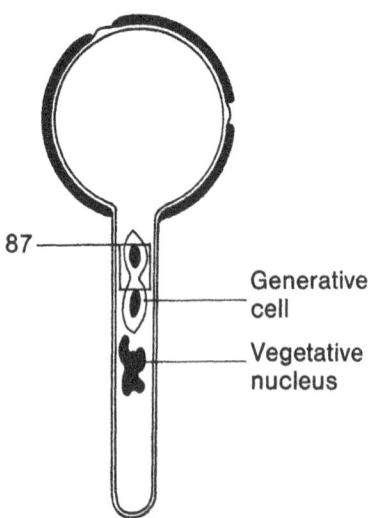

87 ———

Generative
cell

Vegetative
nucleus

Plate 87. Anaphase of the generative cell division in a pollen tube of *Nicotiana tabacum* L.. It shows one of the chromosome groups moving towards one of the poles of the generative cell. Arrows point to the microtubules of the spindle. During anaphase the central region of the generative cell becomes constricted, leading to a dumbbell shape of the dividing generative cell. Arrowheads point to the plasmamembranes in between the vegetative cell and the generative cell. x 115,000.

(Plate 87 courtesy F Ciampolini, Siena).

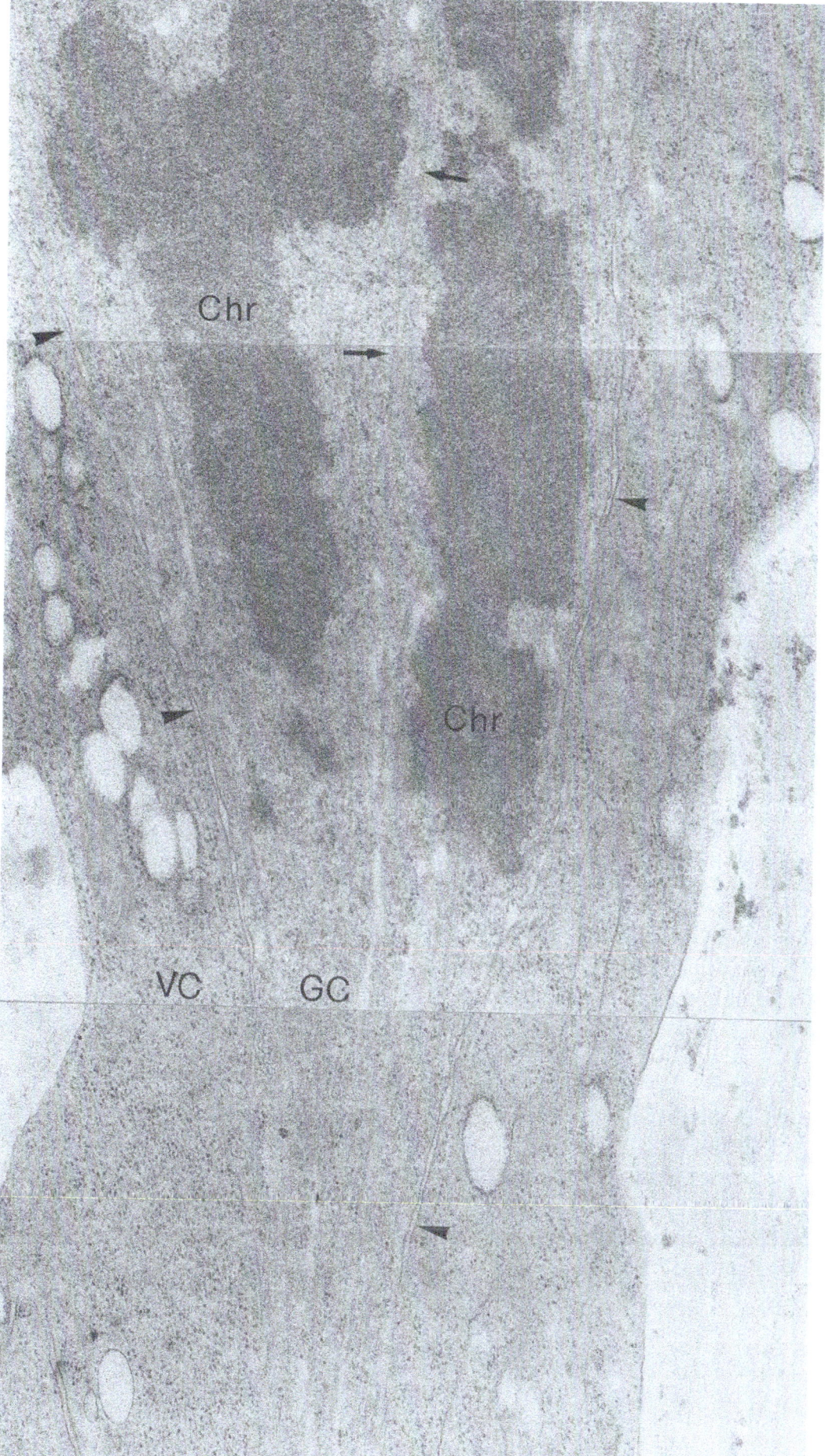

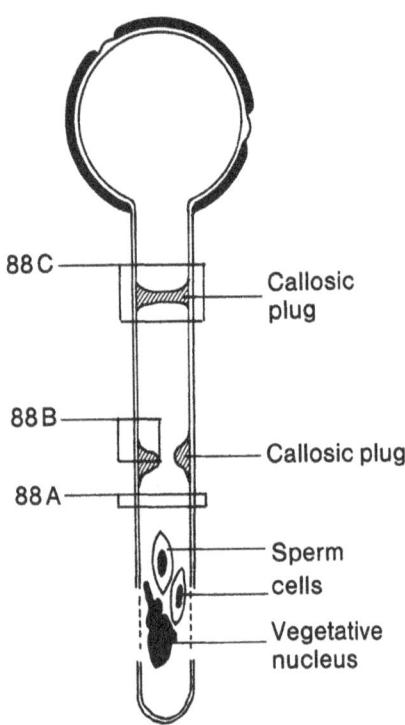

88 C — Callosic plug

88 B — Callosic plug

88 A —

Sperm cells

Vegetative nucleus

Plate 88A. Transverse section through the vacuolated region of a *Petunia hybrida* Hort. ex Vilm. pollen tube. In this region the pollen tube has a thick and rigid inner callosic wall layer (arrows). x 33,000.

Plate 88B. Degeneration of cytoplasm and formation of callose particles in the transition zone between the living and dead part of a pollen tube of *Petunia hybrida* Hort. ex Vilm. The callose particles start to fuse, which results in the formation of a callosic plug. The arrows point to the callosic, inner layer of the pollen tube wall. x 48,000.

Plate 88C. Callose plug in a *Petunia hybrida* Hort. ex Vilm. pollen tube. x 13,000.

(Plates 88A-C reproduced by permission from Cresti M, Van Went J (1976) Planta 133: 35-40).

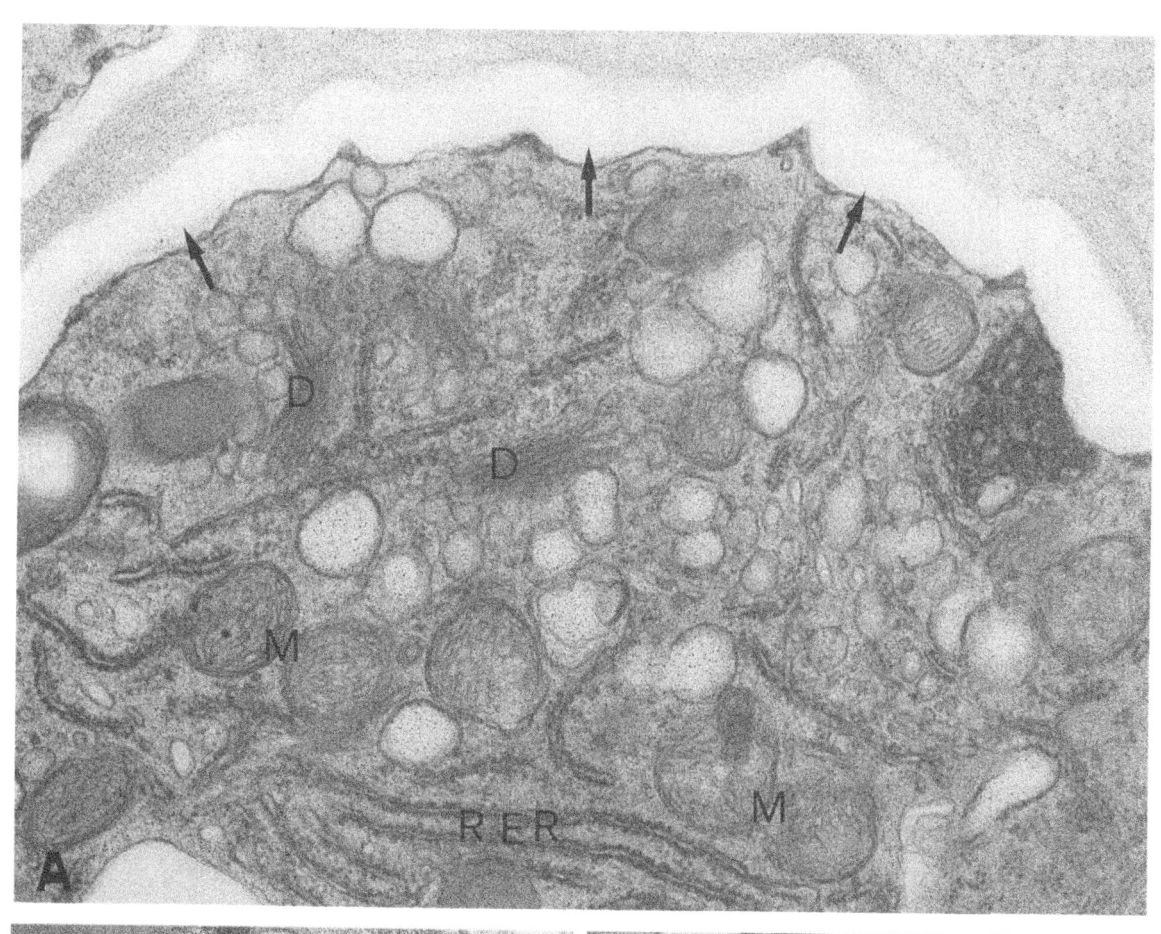

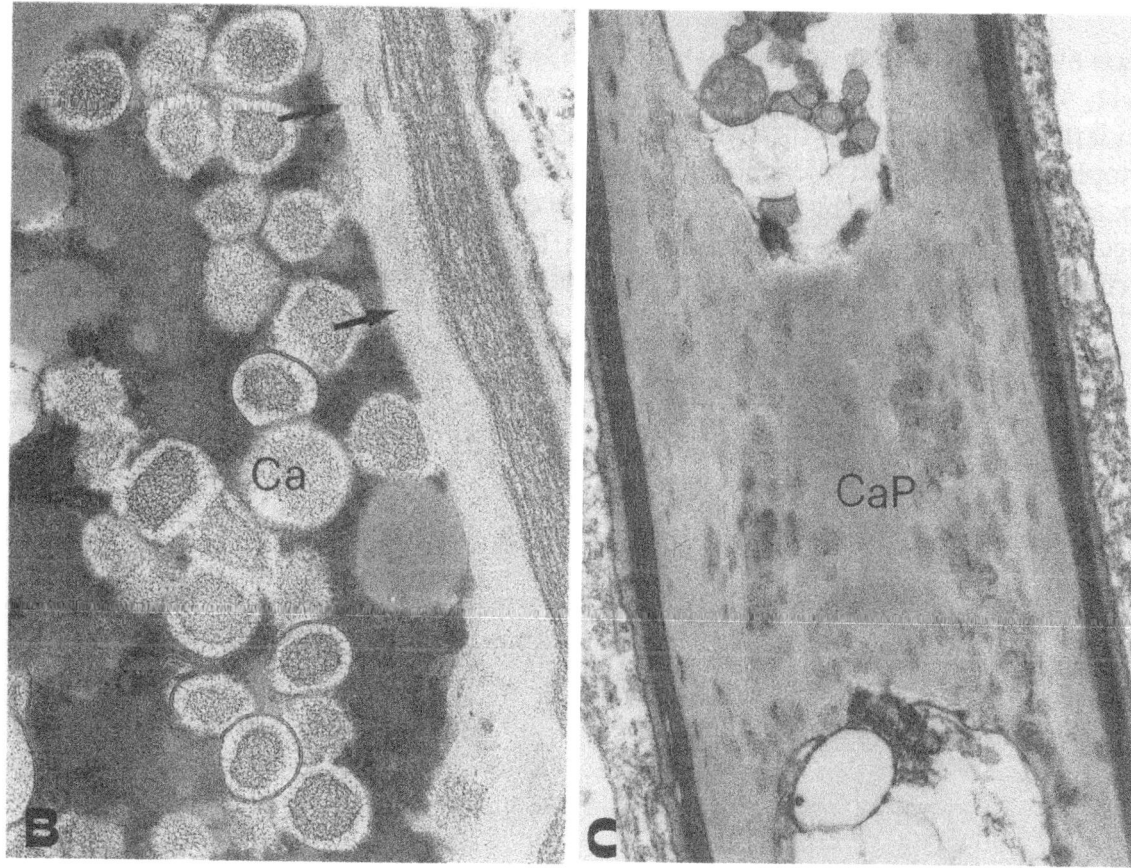

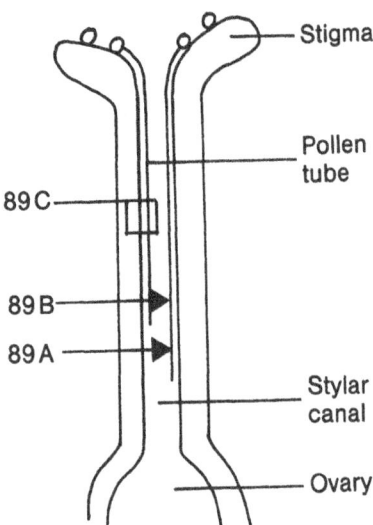

Plate 89A, B. Pollen tubes of *Lilium longiflorum* Thunb. in the stylar canal. The pollen tubes grow over the surface of the stylar canal cells. **A.** shows the apical portion of two tubes. **B.** shows the subapical portion of these tubes. x 660.

Plate 89C. Cross-section of a pollinated *Lilium longiflorum* Thunb. style, showing the stylar canal filled with pollen tubes. The epidermis cells bordering the stylar canal are densely filled with cytoplasm. Some of the pollen tubes have been sectioned near their tip, as can be recognized by the presence of only cytoplasm without vacuoles (arrows). Other pollen tubes have been sectioned at further distance from the tip, indicated by the presence of vacuoles. x 7,200.

(Plates 89A,B courtesy J Janson, Wageningen; Plate 89C courtesy C Keijzer, Wageningen).

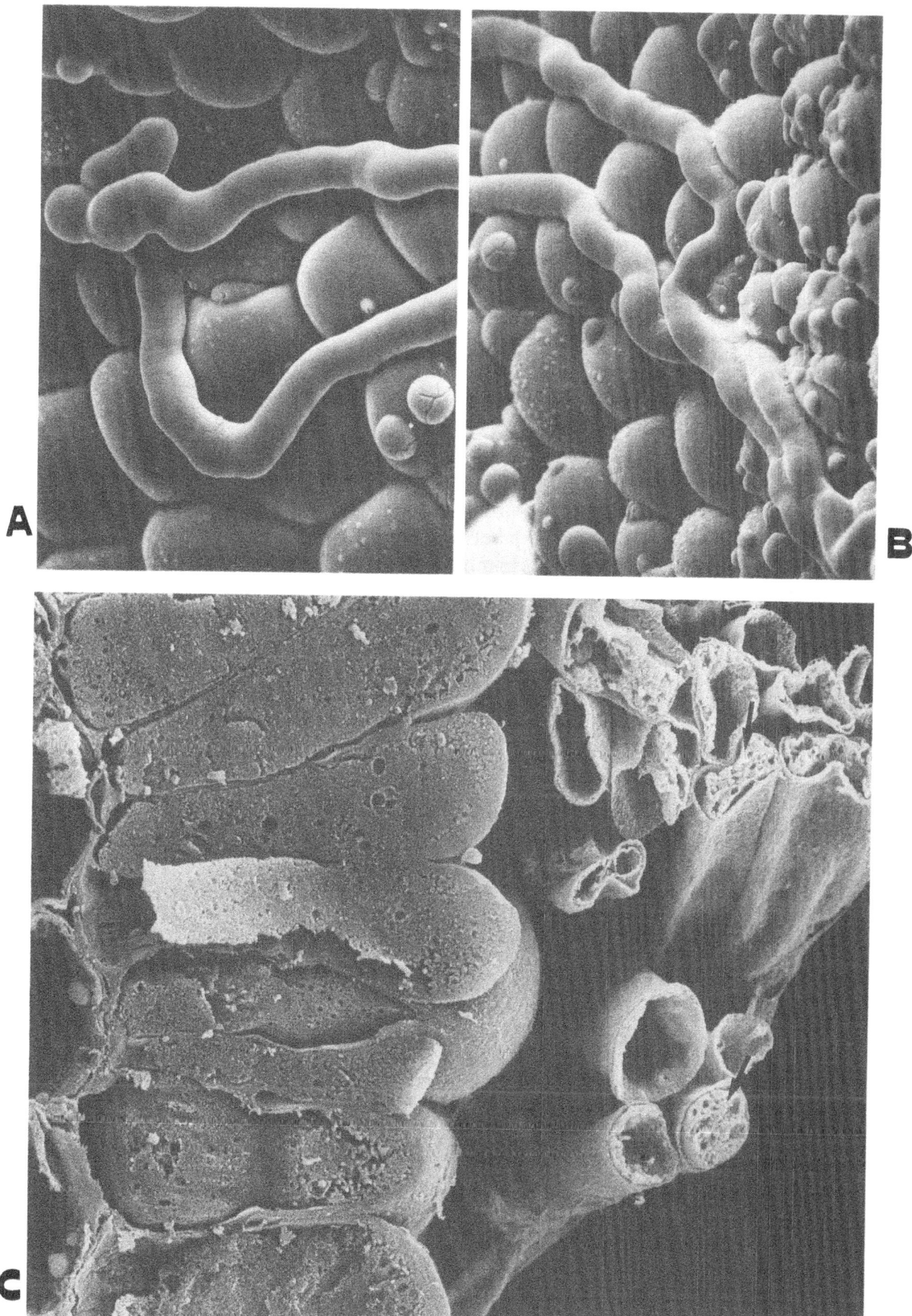

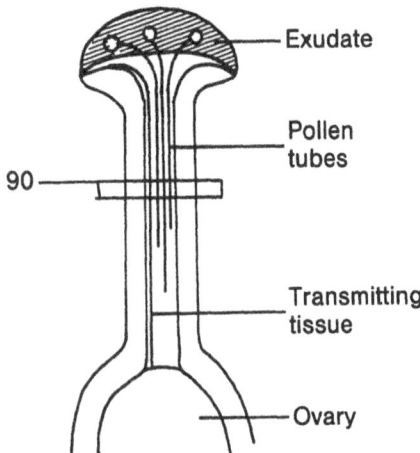

Plate 90. Transverse section of a pollinated style of *Lycopersicon peruvianum* (L.) Mill. The pollen tubes grow down the style through the intercellular substance (arrows) between the cells of the transmitting tissue. The pollen tubes can be easily recognized by the specific structure of their cytoplasm and wall. x 26,400.

(Original M Cresti).

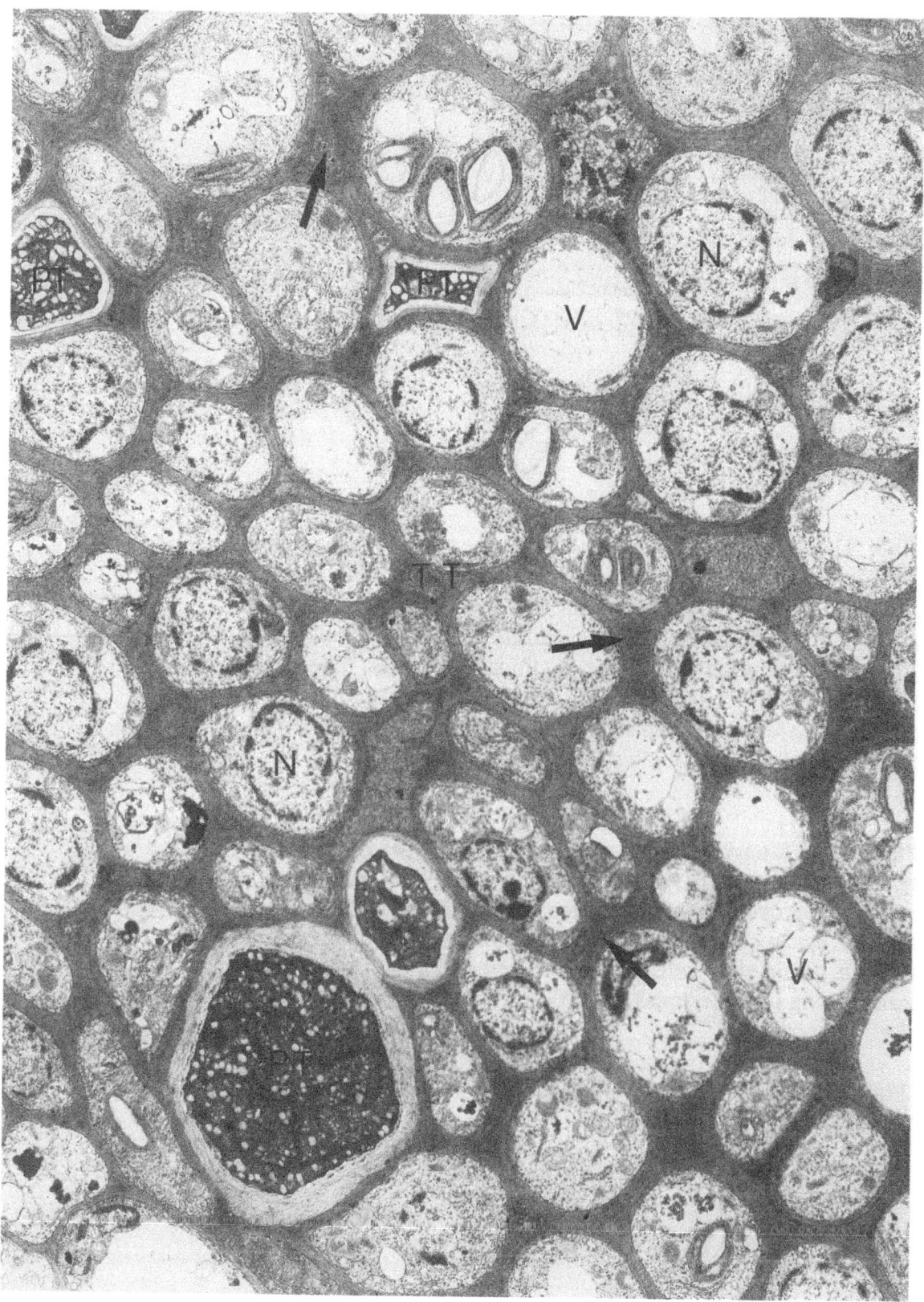

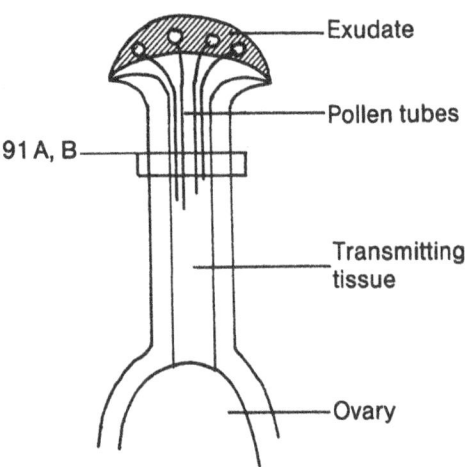

Plate 91A. Transverse section through a style of *Lycopersicon peruvianum* (L.) Mill., which was pollinated with incompatible pollen. The interaction between the pollen tubes and transmitting tissue in the upper half of the style leads to cessation of pollen tube growth. In the cytoplasm of the incompatible pollen tube, large accumulations of concentrically arranged rough endoplasmic reticulum cisterns are present. The inner, callosic wall layer is much thicker than in the compatible pollen tube. x 24,000.

Plate 91B. Transverse section through a style of *Lycopersicon peruvianum* (L.) Mill., which was pollinated with incompatible pollen. After cessation of pollen tube growth, the pollen tubes burst at their apices. The tube content containing large quantities of callosic particles is released in the intercellular space of the transmitting tissue (asterisk). x 7,900.

(Plates 91A, B reproduced by permission from Nettancourt D de et al (1973) J Cell Sci 12: 403-419).

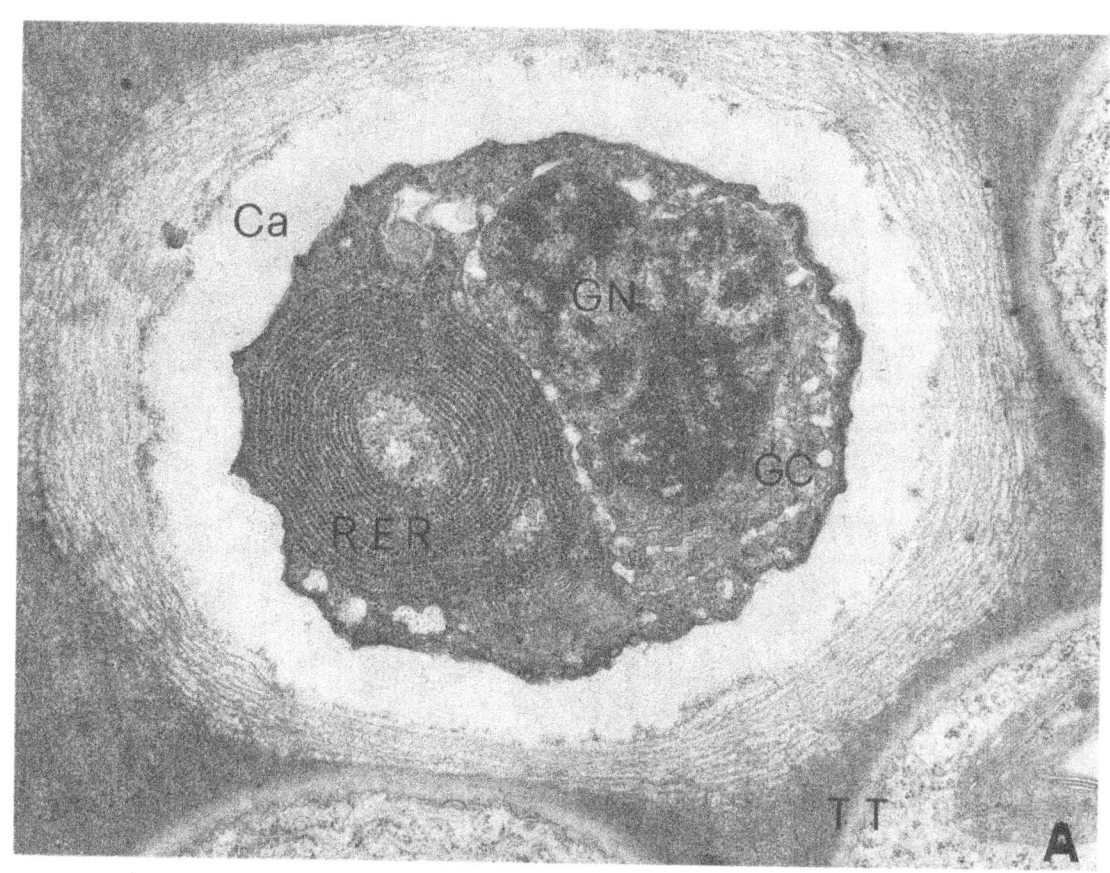

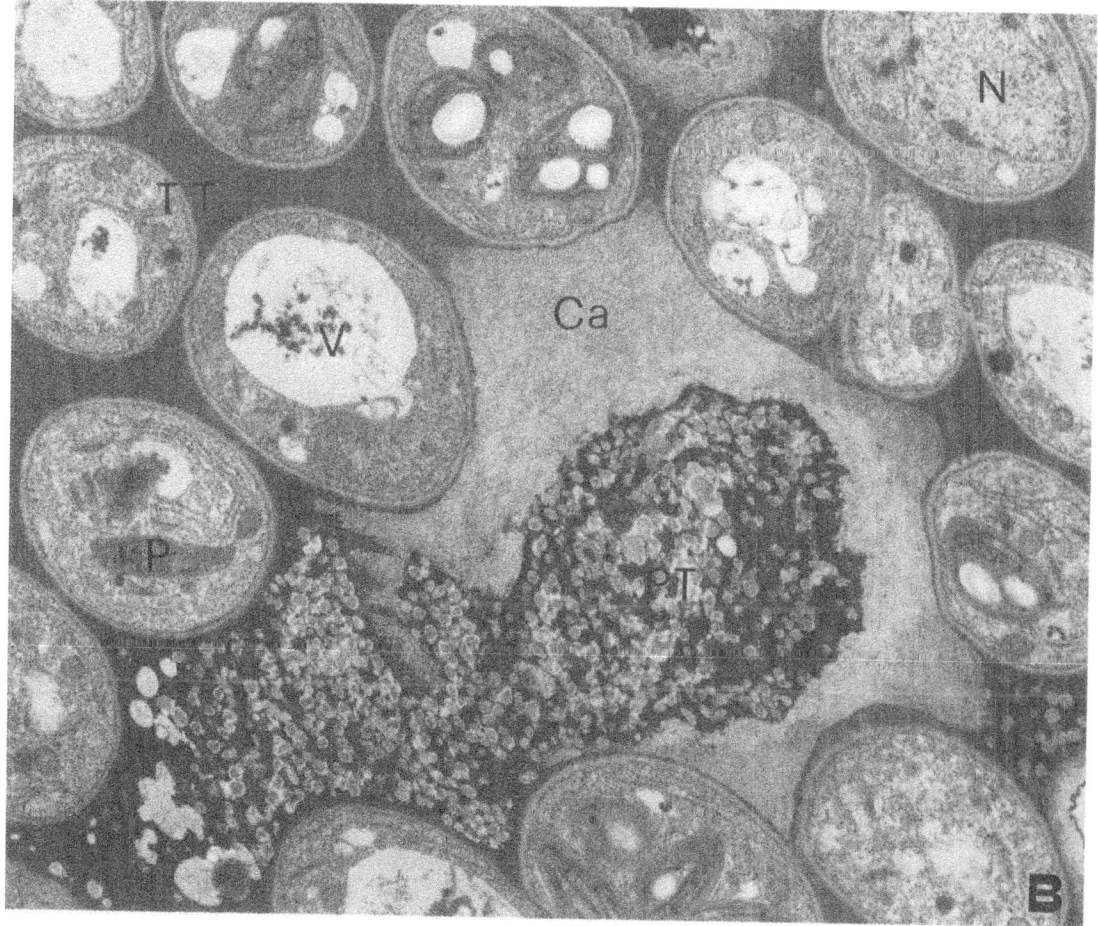

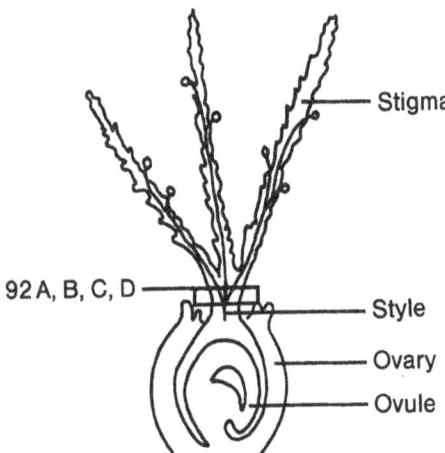

Plate 92A. Transverse section through a pollinated style of *Spinacia oleracea* L.. The style of *Spinacia* is very short, and composed of tightly packed cylindrical cells. The pollen tubes can follow various pathways when they grow towards the ovary. x 3,200.

B. Pollen tube growing through an intercellular space, while at the same time the neighbouring cells are pushed aside. x 13,500.

C. Pollen tube growing in between the cell wall and the plasma membrane (arrows) of a stylar cell (asterisk). Apparently the pollen tubes can pass the cell wall. x 20,000.

D. Pollen tube growing inside the lumen of a stylar cell (asterisk) which has itself degenerated. x 16,000.

(Plates 92A, C courtesy H Wilms, Wageningen; Plates 92B, D reproduced by permission from H Wilms (1980) Acta Bot Neerl 29: 33-47).

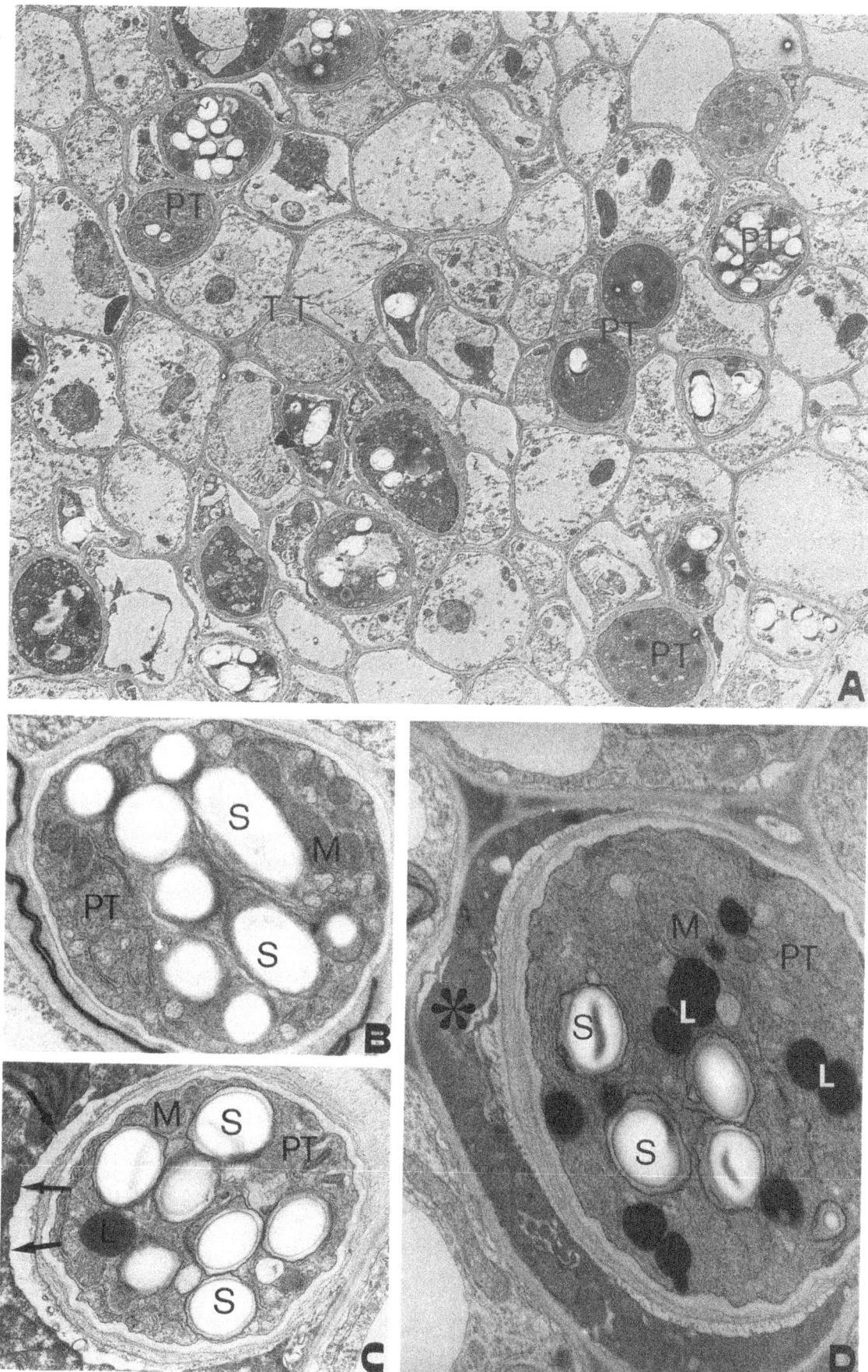

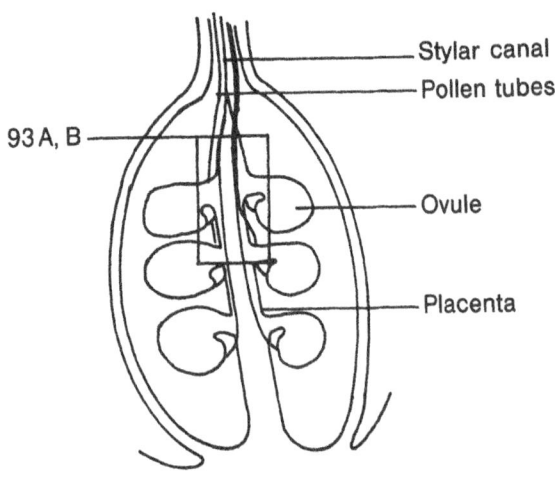

Stylar canal
Pollen tubes

93 A, B

Ovule

Placenta

Plate 93A. Scanning electron micrograph of the placenta of *Lilium longiflorum* Thunb. with pollen tubes. Two ovules have been removed to reveal the pathway of the pollen tubes. F. indicates the attachment sites of the funicle. The pollen tubes grow over the placental ridges in between the ovules. x 300.

Plate 93B. Scanning electron micrograph of fertilized ovules of *Lilium longiflorum* Thunb. The micropyle of the anatropous ovule of lily is positioned very close to the placenta. A pollen tube (arrow) which grows over the placenta continues its growth along the inner integument and enters the micropyle. x 600.

(Plate 93A courtesy C Keijzer, Wageningen; Plate 93B courtesy J Janson, Wageningen).

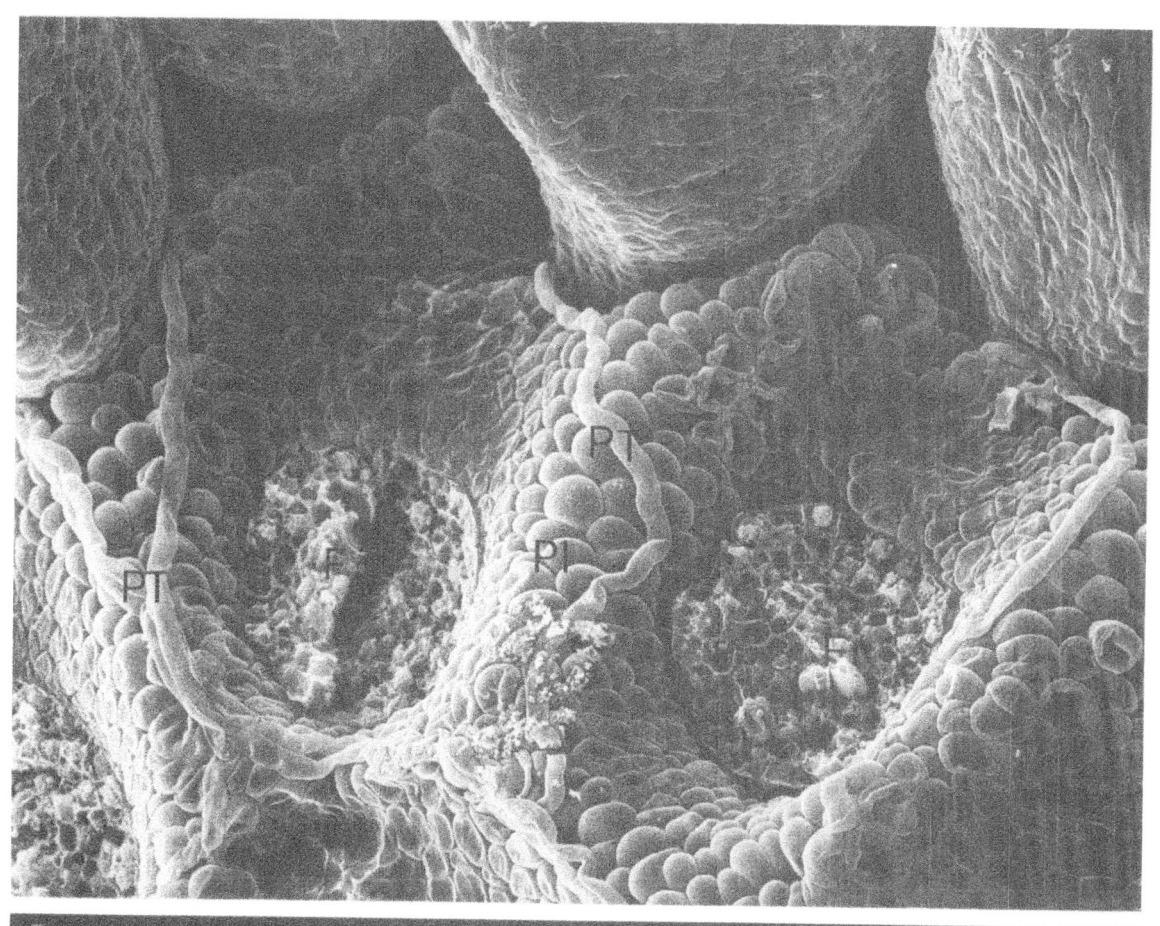

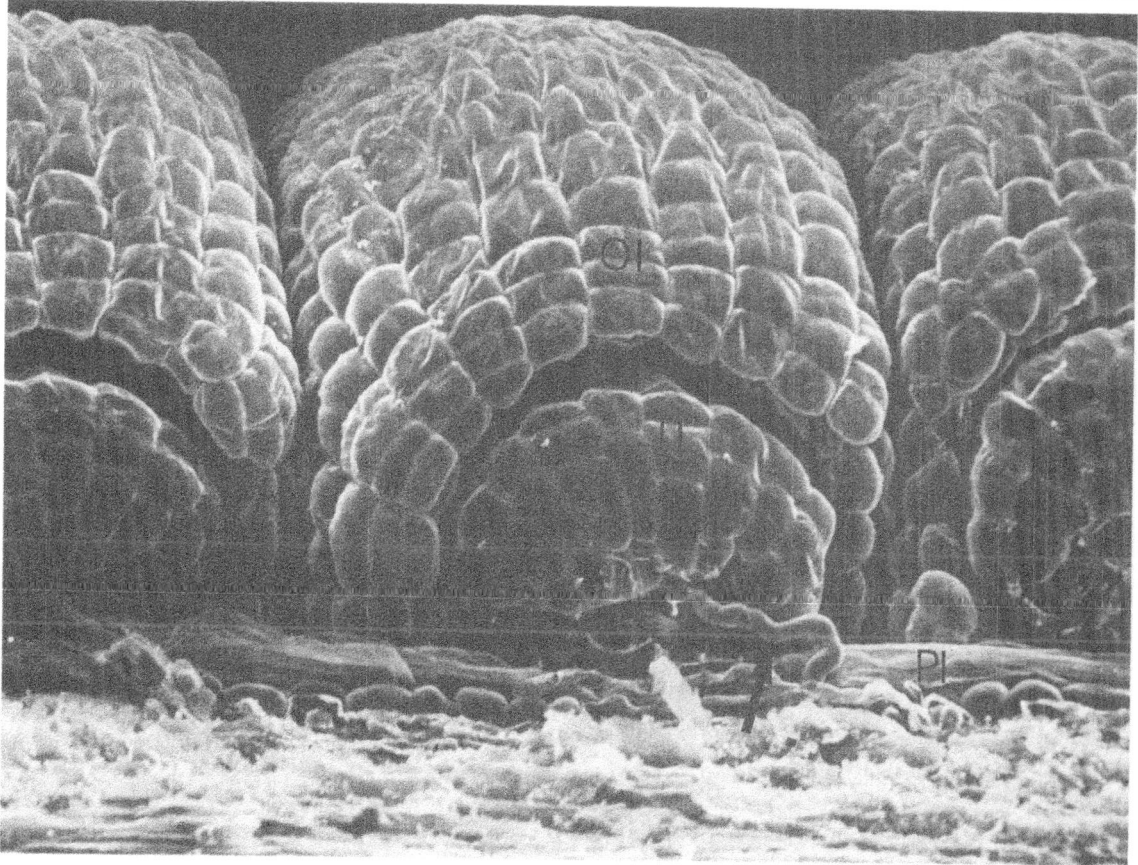

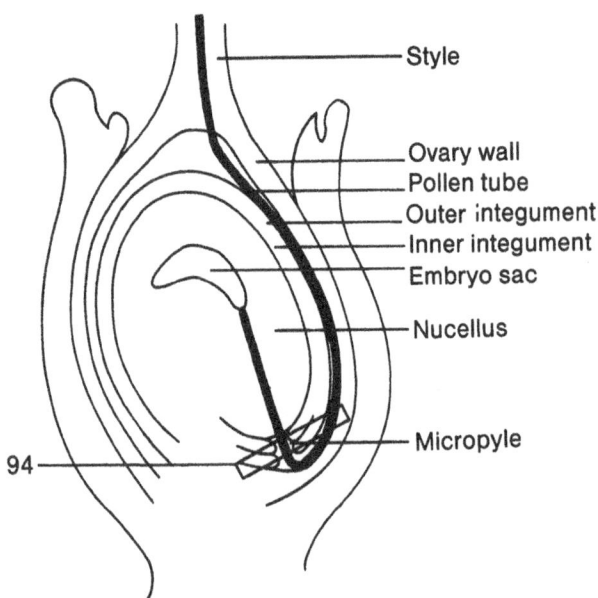

Style

Ovary wall
Pollen tube
Outer integument
Inner integument
Embryo sac

Nucellus

Micropyle

94

Plate 94. Transverse section through the micropylar region of the ovule of *Spinacia oleracea* L.. The ovary of *Spinacia* contains only one bitegmic ovule. The micrograph shows the outer and inner integument. Many transverse sections of pollen tubes can be seen on the outer surface of the ovule, in between the two integuments and within the micropyle. x 1,800.

(Plate 94 reproduced by permission from H Wilms (1981b) Acta Bot Neerl 30: 101-122).

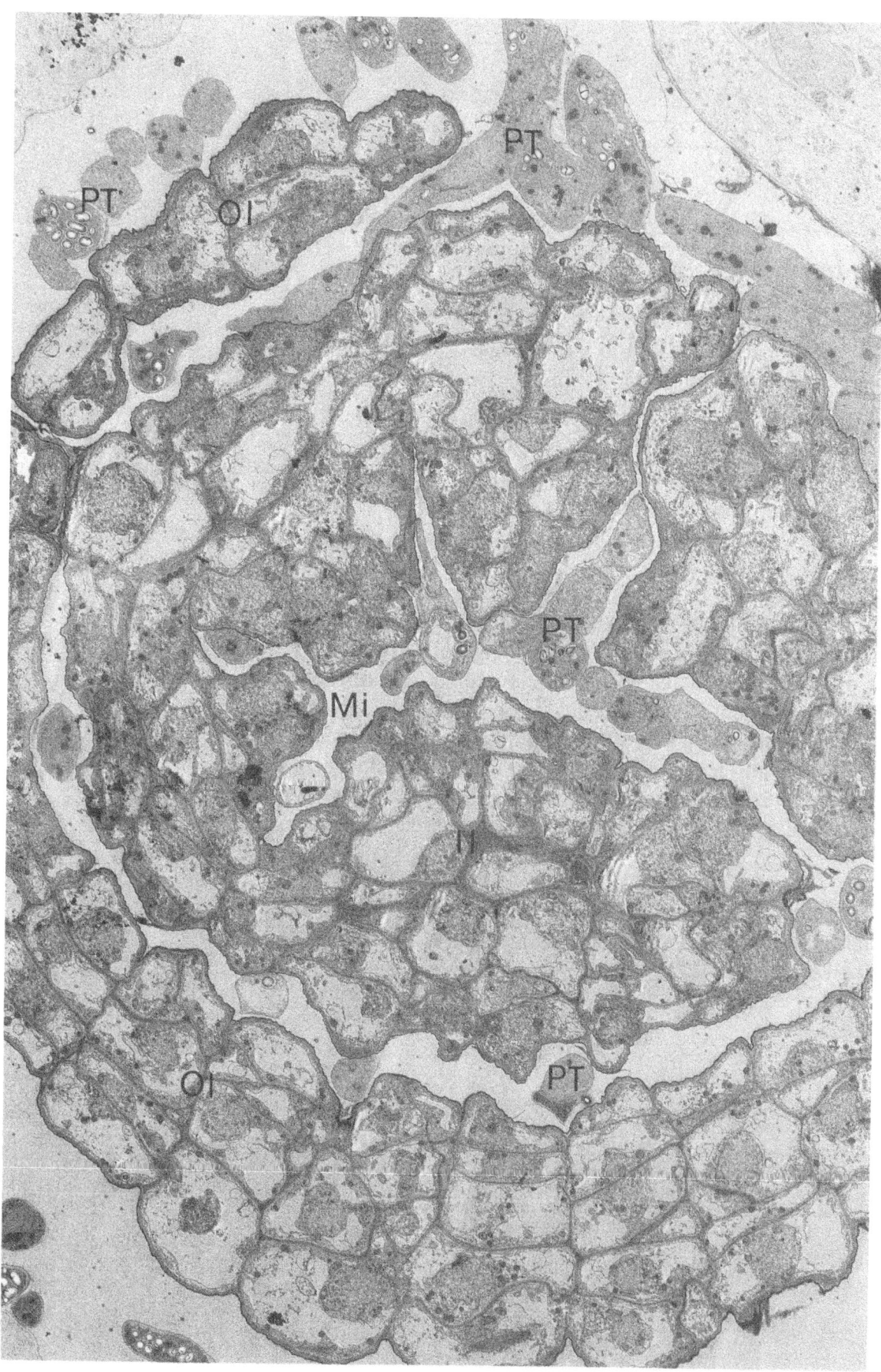

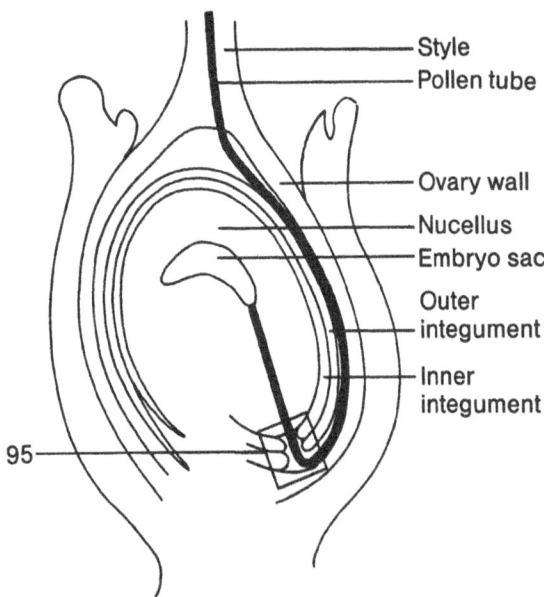

Style
Pollen tube

Ovary wall
Nucellus
Embryo sac
Outer
integument
Inner
integument

95

Plate 95. Longitudinal section through the micropylar region of the ovule of *Spinacia oleracea* L.. Whereas many pollen tubes can reach the ovule, only few actually enter the micropyle and of these only one ultimately enters the nucellus tissue and reaches the embryo sac. The micrograph shows a longitudinal section of one of the pollen tubes inside the micropyle. The pollen tube apparently do not enter into the nucellus tissue directly, but initially grow over the surface of the nucellus tissue, and can even form branches. x 3,100.

(Plate 95 reproduced by permission from H Wilms (1981b) Acta Bot Neerl 30: 101-122).

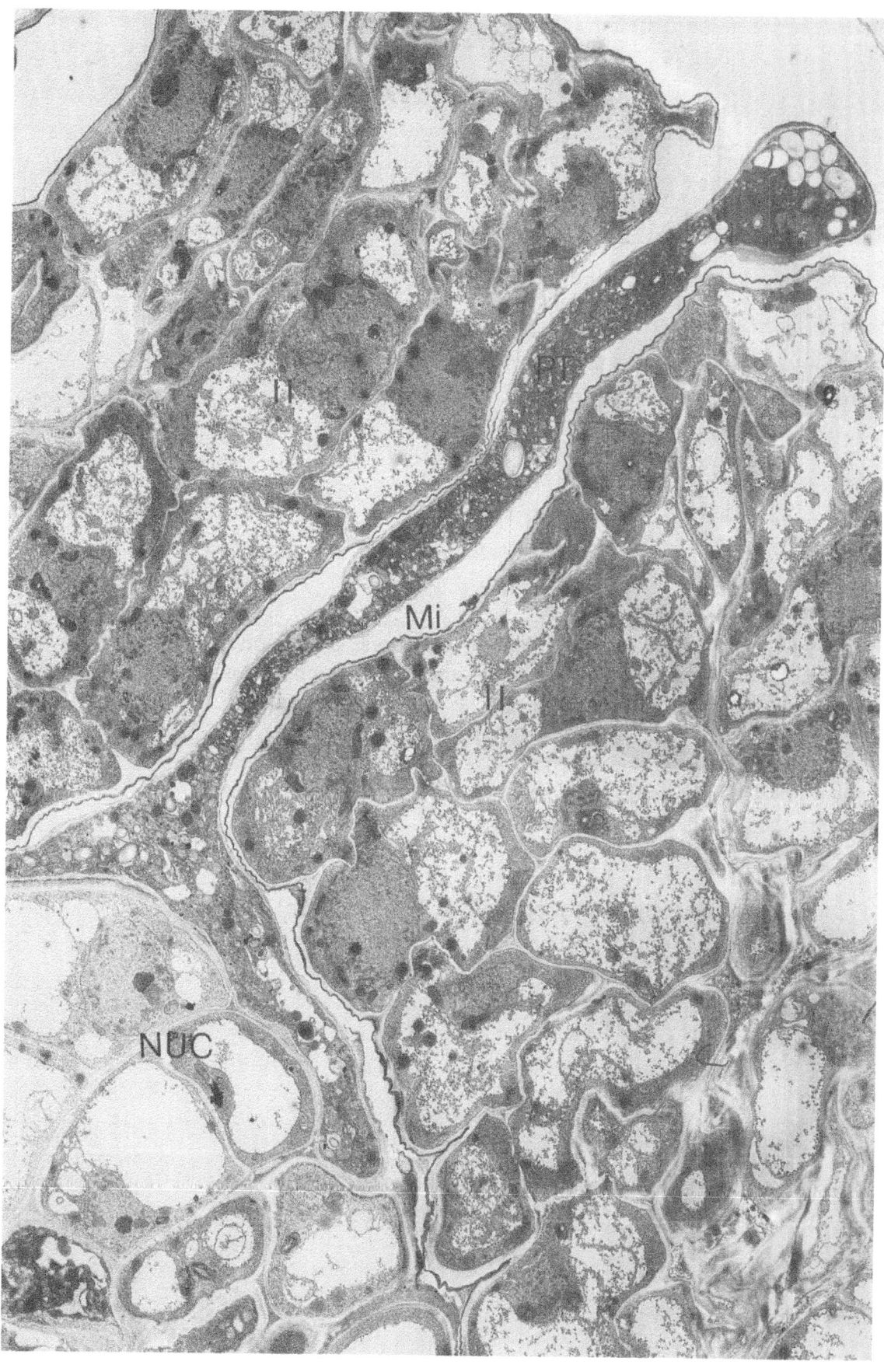

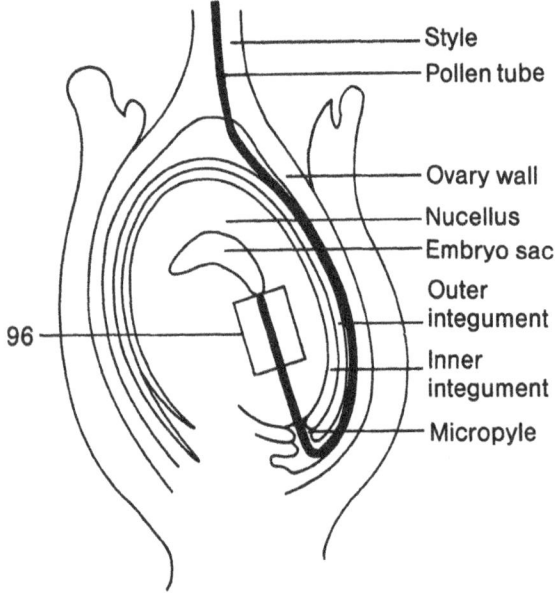

Style
Pollen tube

Ovary wall
Nucellus
Embryo sac
Outer integument

96

Inner integument
Micropyle

Plate 96. Longitudinal section through the ovule of *Spinacia oleracea* L.. The ovule of *Spinacia* is crassinucellate, which means that at maturity stage, the embryo sac is still surrounded by a nucellar tissue. To reach the embryo sac from the micropyle the pollen tube has to pass this nucellar tissue. The pollen tube forces its way towards the embryo sac through the middle lamellae in between the nucellus cells. x 2,900.

(Plate 96 reproduced by permission from H Wilms (1981b) Acta Bot Neerl 30: 101-122).

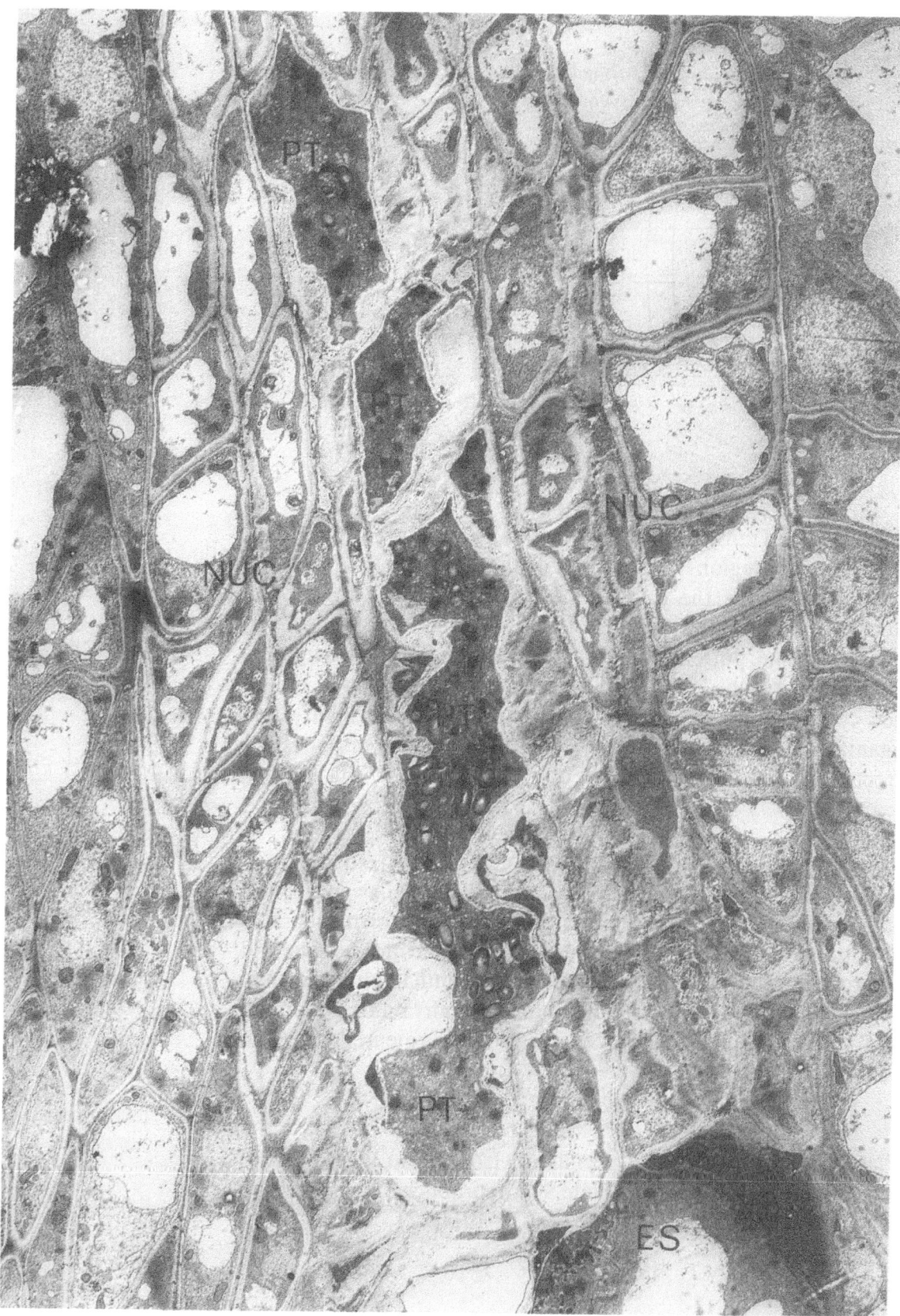

229

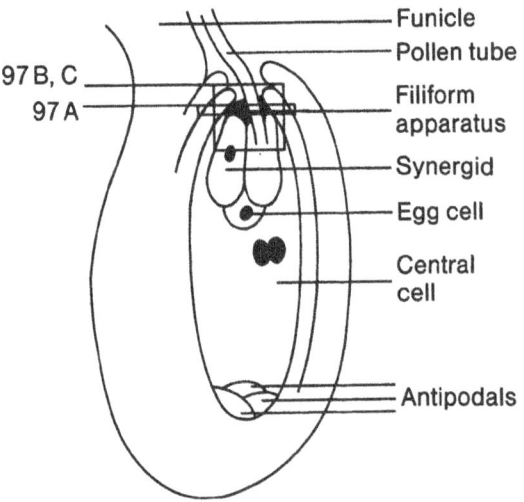

97 B, C
97 A

Funicle
Pollen tube
Filiform
apparatus
Synergid
Egg cell
Central
cell
Antipodals

Plate 97 A, B. Transverse (**A**) and longitudinal (**B**) section through the micropylar region of a fertilized ovule of *Petunia hybrida* Hort. ex Vilm., showing the passage of the pollen tube through the filiform apparatus and the pollen tube entrance into one of the synergids. The pollen tube ruptures as soon as the synergid cytoplasm is reached (arrow). The subsequent injection of pollen tube cytoplasm caused drastic changes of the ultrastructure and organization of the penetrated synergid. The second synergid is not affected. x 5,500.

Plate 97C. Longitudinal section through a fertilized ovule of *Spinacia oleracea* L.. After its arrival at the micropylar pole of the embryo sac, the pollen tube grows through the filiform apparatus into one of the synergids. Soon after its entrance into the synergid the growth of the pollen tube ceases, and a terminal pore is formed or the pollen tube ruptures. Subsequently a portion of the pollen tube content, containing the sperm cells, is released into the synergid. x 5,800.

(Plates 97A,B reproduced by permission from J Van Went (1970b) Acta Bot Neerl 19: 468-480; Plate 97C reproduced by permission from H Wilms (1981) Acta Bot Neerl 30: 101-122).

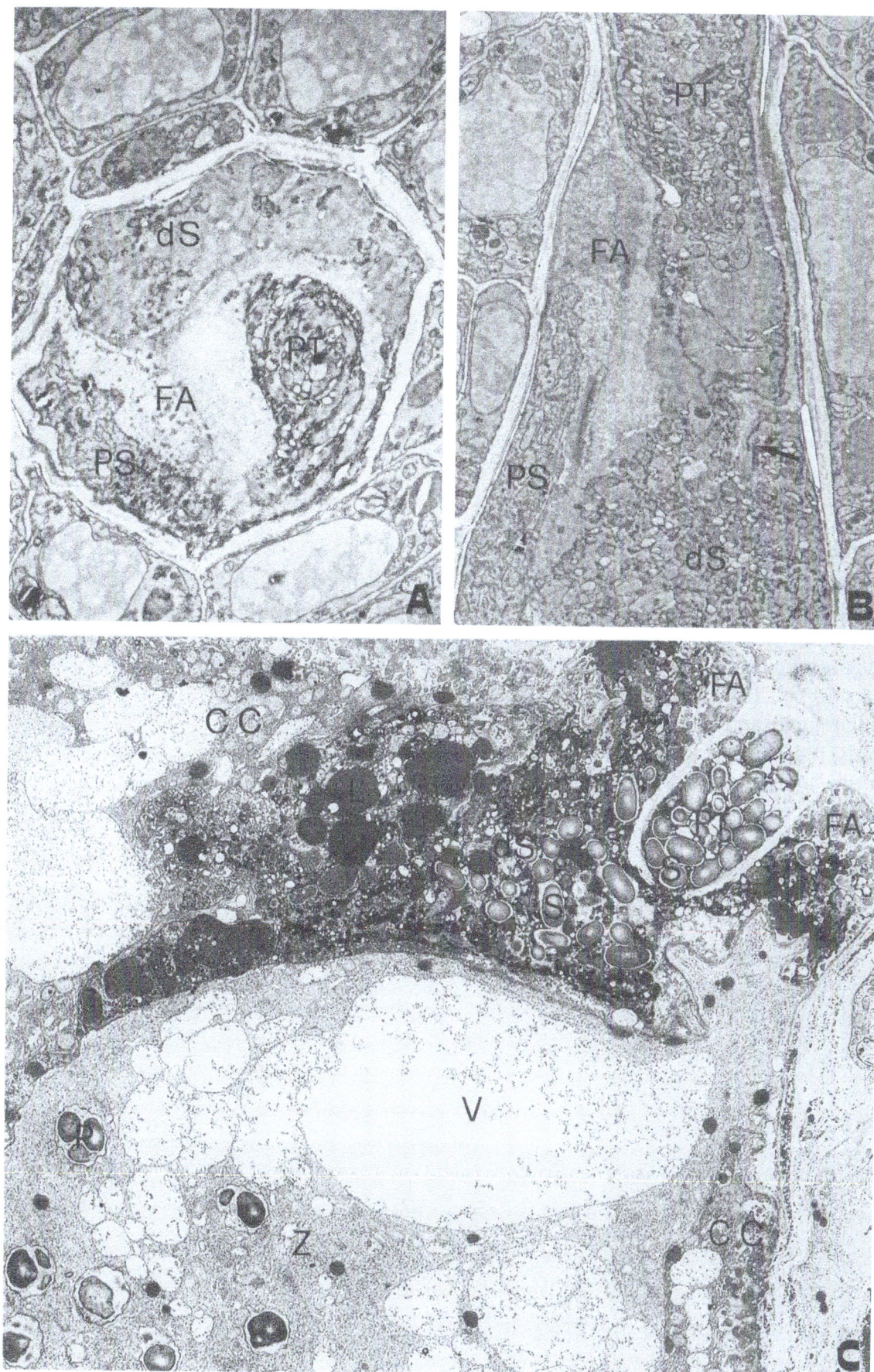

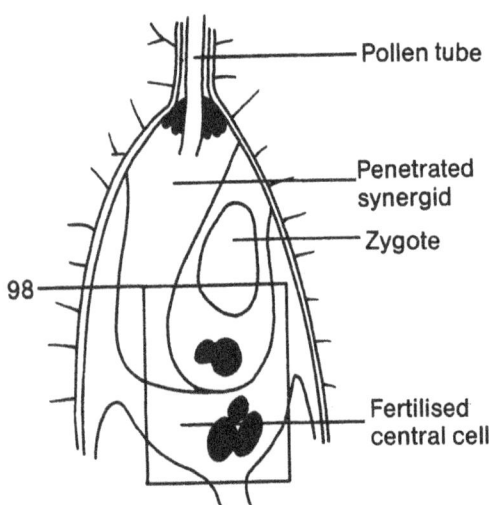

Pollen tube

Penetrated
synergid

Zygote

98

Fertilised
central cell

Plate 98. Longitudinal section through a fertilized ovule of *Spinacia oleracea* L., showing the zygote and fertilized central cell. In the zygote nuclear fusion is already completed, although the positions of the former male nucleus can still be traced by the shape of the zygote nucleus and the position of the smaller nucleolus. In the central cell (endosperm cell) the fusion of the three nuclei (two polar nuclei and male nucleus) is still in process. In between the two cells, remnants of the penetrated synergid, indicating the pathway of the sperm cells, are still present (indicated by arrows). x 3,600.

(Plate 98 reproduced by permission from H Wilms (1981b) Acta Bot Neerl 30: 101-122).

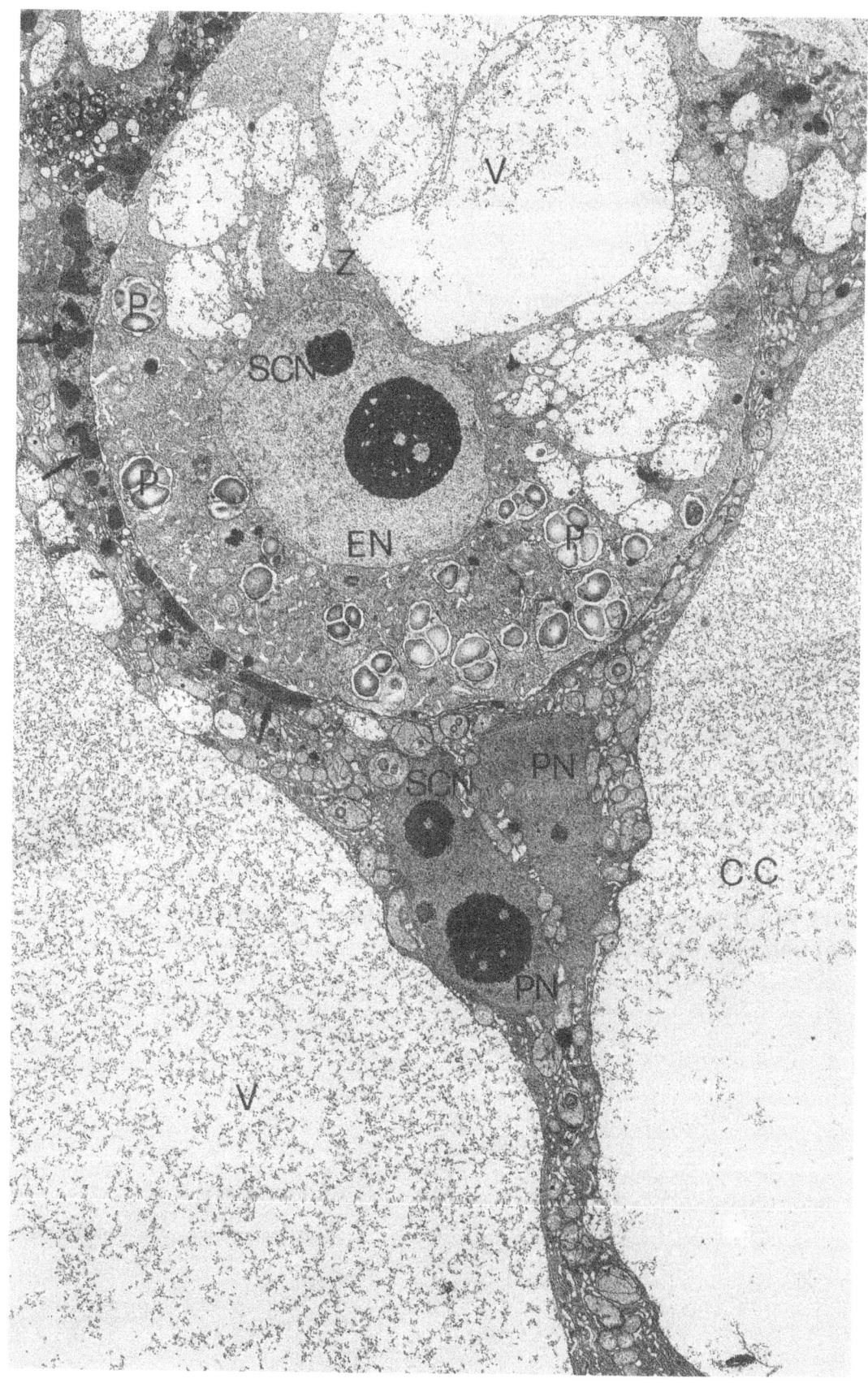

233

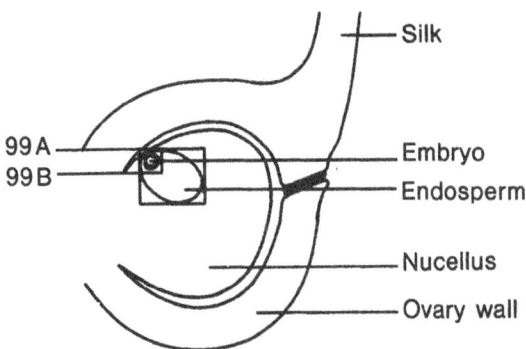

Plate 99A. Longitudinal section through a fertilized ovule of *Zea mays* L., showing an early stage of endosperm and embryo development. x 200.

Plate 99B. Longitudinal section through a fertilized ovule of *Zea mays* L., showing an early development stage of the proembryo. The insert shows a light-optical overview of the cellular endosperm and the position of the young embryo. The first divisions are unequal, which results in a pro-embryo, composed of small, cytoplasmic-rich apical cells (asterisks), and a large highly vacuoled basal cells. The apical cells develop into the embryo proper, while the basal cells will form the suspensor. Note that the synergid which was penetrated by the pollen tube is still present. x 2,000.

(Plates 99A,B reproduced by permission from Schel J, Kieft H (1986) Can J Bot 64: 2227-2238).

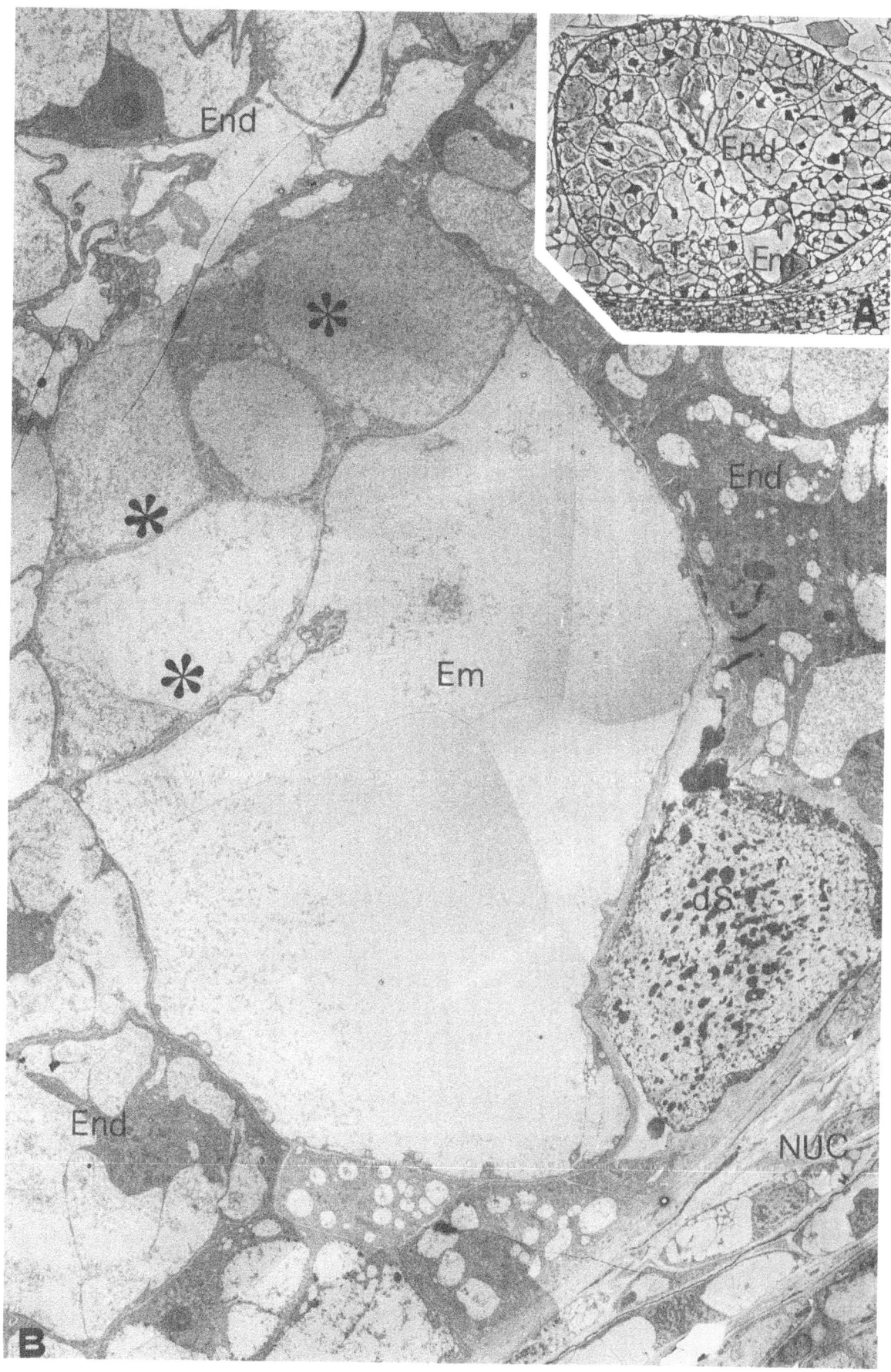

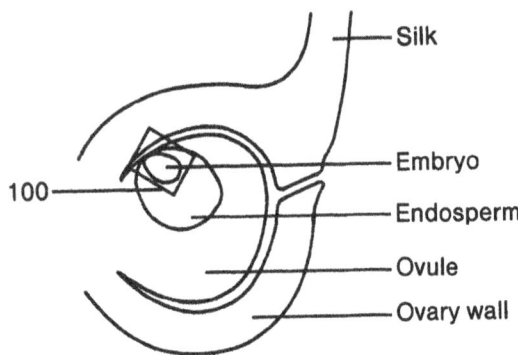

Silk

Embryo

Endosperm

Ovule

Ovary wall

100

Plate 100. Longitudinal section through the fertilized ovule of *Zea mays* L.. The embryo is at the early globular stage. The apical cells (asterisk) are smaller and much richer in cytoplasm then the cells near the suspensor. The embryo is surrounded by cellular endosperm. x 2,000.

(Plate 100 reproduced by permission from Schel J, Kieft H (1986) Can J Bot 64: 2227-2238).

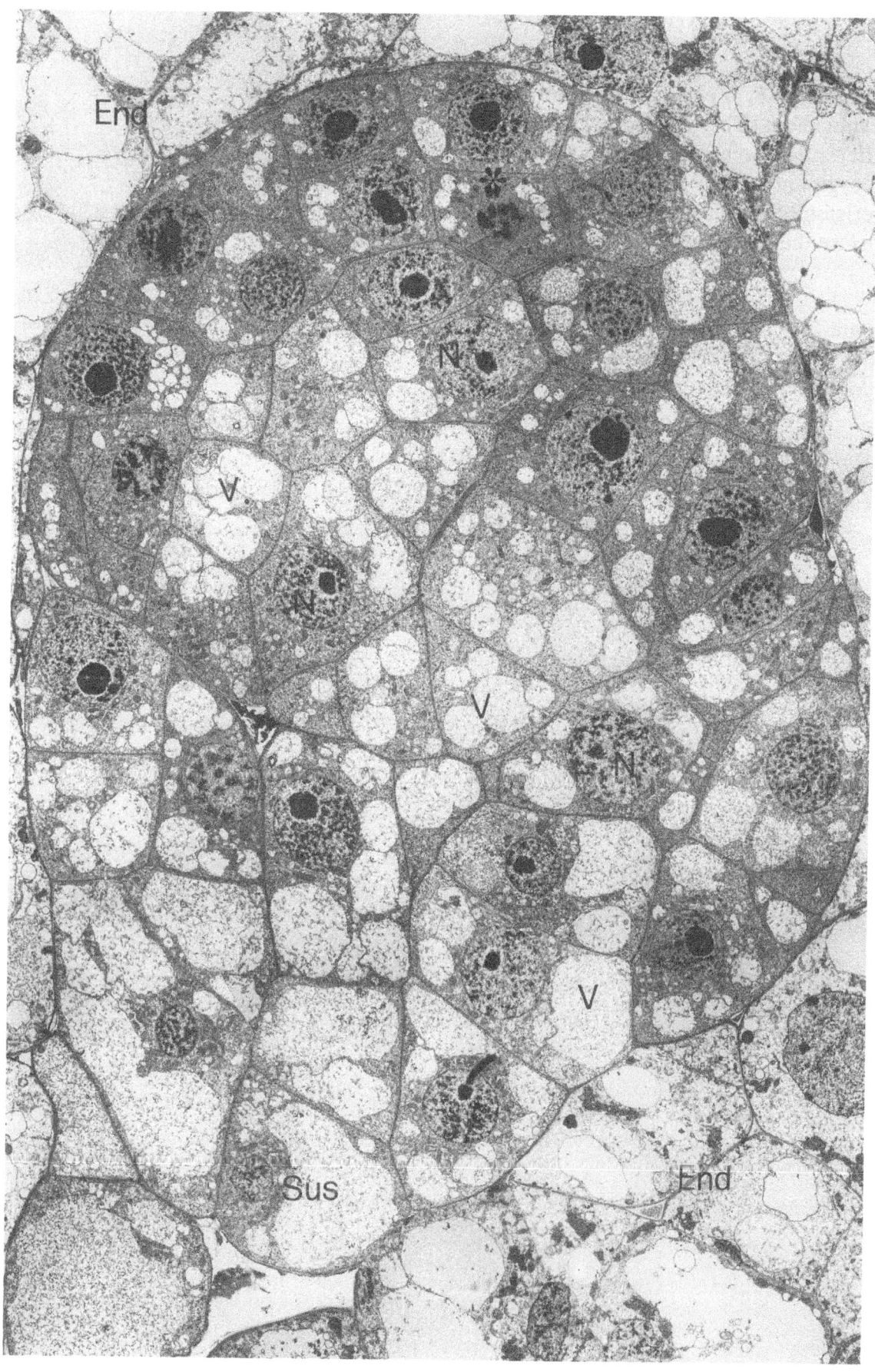

237

Plate 101A-F. *Zea mays* L. embryos at different stages of development. The development from pro-embryo (**A**) to a well differentiated embryo (**F**) is marked by the initiation (**B**) and rapid enlargment of the scutellum and the initiation of the shoot meristem. Subsequently, the coleoptile is formed (**C**) which will enclose the shoot apex. x 530.

(Plate 101 reproduced by permission from A Van Lammeren (1986a) Acta Bot Neerl 35: 169-188).

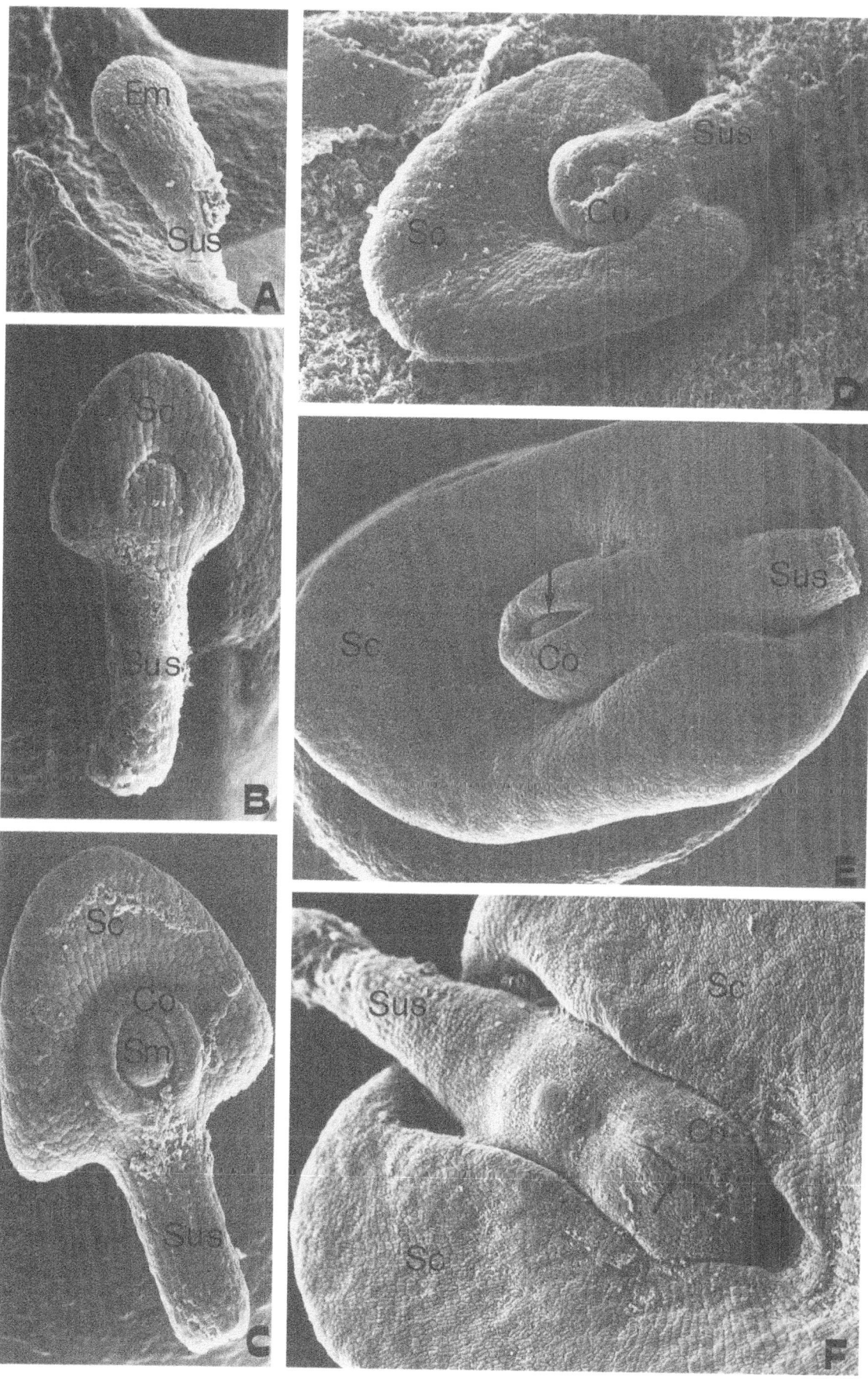

References

Barnes S H, Blackmore S (1988) Pollen ontogeny in *Catananche caerulea* L. (Compositae: Lactucaceae) II. Free microspore stage to the formation of male germ unit. Ann Bot 62: 615-623

Blackmore S, Barnes S H (1988) Pollen ontogeny in *Catananche caerulea* L. (Compositae: Lactucaceae) I: Premeiotic phase to establishment of tetrads. Ann Bot 62: 605-614

Charzynska M, Ciampolini F, Cresti M (1988) Generative cell division and sperm cell formation in barley. Sex Plant Reprod 1: 240-248

Charzynska M, Murgia M, Cresti M (1990) Microspore of *Secale cereale* as a transfer cell type: Protoplasma 158: 26-32

Charzynska M, Murgia M, Milanesi C, Cresti M (1989) Origin of the sperm cell association in the "male germ unit" of *Brassica* pollen. Protoplasma 149: 1-4

Ciampolini F, Cresti M (1981) Atlante dei principali pollini allergenici presenti in Italia. Università di Siena, Siena

Ciampolini F, Cresti M, Kapil R N (1983) Fine structural and cytochemical characteristics of style and stigma in olive. Caryologia 36: 211-230

Ciampolini F, Cresti M, Sarfatti G, Tiezzi A (1981) Ultrastructure of the stylar canal cells of *Citrus limon* (Rutaceae). Pl Syst Evol 138: 263-274

Cresti M, Ciampolini F, Van Went J L (1991) Strip-Shaped projections at the cytoplasmic face of outer membrane of the generative cell in *Amaryllis bella-donna*. Ann Bot 68: 105-107

Cresti M, Keijzer C J, Tiezzi A, Ciampolini F, Focardi S (1986) Stigma of *Nicotiana*: ultrastructural and biochemical studies. Amer J Bot 73: 1713-1722

Cresti M, Milanesi C, Tiezzi A, Ciampolini F, Moscatelli A (1988) Ultrastructure of *Linaria vulgaris* pollen grains. Acta Bot Neerl 37: 379-386

Cresti M, Van Went J L (1976) Callose deposition and plug formation in *Petunia* pollen tubes in situ. Planta 133: 35-40

De Boer-De Jeu M (1978) Megasporogenesis, a comparative study of the ultrastructural aspects of megasporogenesis in *Lilium*, *Allium* and *Impatiens*. Comm Agricult Univ Wageningen 16: 1-128

Heslop-Harrison J, Heslop-Harrison Y (1989) Myosin associated with the surface of organnelles, vegetative nuclei and generative cells in angiosperm pollen grains and tubes. J Cell Sci 94: 319-325

Keijzer C J (1987) The processes of anther dehiscence and pollen dispersal. II. The formation and the transfer mechanism of pollenkitt, cell-wall development of the loculus tissues and a function of orbicules in pollen dispersal. New Phytol 105: 499-507

Nettancourt de D, Devreux M, Bozzini A, Cresti M, Pacini E, Sarfatti G (1973) Ultrastructural aspect of the self-incompatibility mechanism in *Lycopersicum peruvianum* (L.) Mill.. J Cell Sci 12: 403-149

Schel J H N, Kieft H (1986) An ultrastructural study of embryo and endosperm development during in vitro culture of maize ovaries (*Zea mays*). Can J Bot 64: 2227-2238

Tang X J, Hepler P K, Scordilis S P (1989) Immunochemical and immunocytochemical identification of a myosin heavy chain polypeptide in *Nicotiana* pollen tubes. J Cell Sci 92: 569-574

Theunis C H (1990) Ultrastructural analysis of *Spinacia oleracea* sperm cells isolated from mature pollen grains. Protoplasma 158: 176-181

Theunis C H, Van Went J L (1989) Isolation of sperm cells from mature pollen grains of *Spinacia oleracea* L. Sex Plant Reprod 2: 97-102

Van Aelst A C, Mueller T, Dueggelin M, Guggenheim R (1989) Three-dimensional observations on freeze-fractured frozen hydrated *Papaver dubium* pollen wityh cryo-scanning electron microscopy. Acta Bot Neerl 38: 25-30

Van Lammeren A A M (1986a) Developmental morphology and cytology of the young maize embryo (*Zea mays* L.). Acta Bot Neerl 35: 169-188

Van Lammeren A A M (1986b) A comparative ultrastructural study of the megagametophytes in two strains of *Zea mays* L., before and after fertilization. Agric Univ Wageningen Pap 86-1

Van Went J L (1970a) The ultrastructure of the synergids of *Petunia*. Acta Bot Neerl 19: 121-132

Van Went J L (1970b) The ultrastructure of the fertilized embryo sac of *Petunia*. Acta Bot Neerl 19: 468-480

Van Went J L (1974) The ultrastructure of *Impatiens* pollen. In: Linskens H F (ed) Fertilization in higher plants. North-Holland, Amsterdam, pp 81-88

Van Went J L (1981) Some cytological and ultrastructural aspects of male sterility in *Impatiens*. Acta Soc Bot Pol 50: 249-252

Van Went J L (1984) Unequal distribution of plastids during generative cell formation in *Impatiens*. Theor Appl Genet 68: 305-309

Van Went J, Cresti M (1988a) Pre-fertilization degeneration of both synergids in *Brassica campestris* ovules. Sex Plant Reprod 1: 208-216

Van Went J, Cresti M (1988b) Cytokinesis in microspore mother cells of *Impatiens sultani*. Sex Plant Reprod 1: 228-233

Van Went J, Cresti M (1989) Cytoplasmic differentation during tetrad formation and early microspore development in *Impatiens sultani*. Protoplasma 148: 1-7.

Van Went J L, Kwee H S (1990) Enzymatic isolation of living embryo sacs of *Petunia*. Sex Plant Reprod 3: 257-262

Wilms H J (1980) Ultrastructure of the stigma and style of spinach in relation to pollen germination and pollen tube growth. Acta Bot Neerl 29: 33-47

Wilms H J (1981a) Ultrastructure of the developing embryo sac of spinach. Acta Bot Neerl 30: 75-99

Wilms H J (1981b) Pollen tube penetration and fertilization in spinach. Acta Bot Neerl 30: 101-122

Impatiens sultani Hook f.	plate 2B, 3C, 6A, 6B, 6C, 7B, 8B, 9A, 9B, 30B
Indigofera polysphaera Baker	plate 37D
Ipomoea purpurea Roth.	plate 56A, 56B, 68B
Iris florentina L.	plate 69C, 69D
Lilium longiflorum Thunb.	plate 67A, 67B, 69A, 69B, 89A, 89B, 89C, 93B
Linaria vulgaris Mill.	plate 34A, 70A
Lycopersicon esculentum (L.) Mill.	plate 62B, 93A
Lycopersicon peruvianum (L.) Mill.	plate 62A, 63, 70B, 72, 73B, 74, 90, 91A, 91B
Malus communis Poiret	plate 64
Nicotiana alata Link & Otto	plate 25B, 27A, 27B, 34, 55A, 55B, 55C, 61A, 61B, 78, 79, 81, 82, 83, 84H, 84I, 84K, 84L
Nicotiana tabacum L.	plate 59A, 59B, 60, 73A, 80, 85A, 85B, 86A, 86B, 86C, 87
Olea europaea L.	plate 13A, 13B, 65
Papaver dubium L.	plate 26A, 26B, 26C, 26D
Papaver rhoeas L.	plate 28B
Parietaria judaica L.	plate 13D
Passiflora racemosa Brot.	plate 38A
Passiflora vespertilio L.	plate 38B, 38D
Petunia hybrida Hort. ex Vilm.	plate 46B, 54, 88A, 88B, 88C, 97A, 97B
Picea abies(L.) Karsten	plate 36C
Plumbago zeylanica L.	plate 55D, 55E
Psoralea bituminosa L.	plate 37C
Secale cereale L.	plate 22A, 22B
Spinacia oleracea L.	plate 33B, 33C, 33D, 43, 44, 46C, 49, 50A, 50B, 51A, 51B, 52A, 75A, 75B, 92A, 92B, 92C, 92D, 94, 95, 96, 97C, 98
Tradescantia virginiana L.	plate 23
Zea mays L.	plate 52B, 52C, 99, 100, 101

Subject index

The manufacturer's authorised representative in the EU is Springer
Nature Customer Service Centre GmbH, Europaplatz 3, 69115 Heidelberg,
Germany. If you have any concerns regarding our products, please
contact ProductSafety@springernature.com

Printed and bound by CPI Group (UK) Ltd, Croydon, CR0 4YY
28/04/2026
02098548-0001